Managing Electric Vehicle Power

Managing Electric Vehicle Power

Warrendale, Pennsylvania, USA

400 Commonwealth Drive
Warrendale, PA 15096-0001 USA
E-mail: CustomerService@sae.org
Phone: 877-606-7323 (inside USA and Canada)
724-776-4970 (outside USA)
FAX: 724-776-0790

Library of Congress Catalog Number 2020936333
http://dx.doi.org/10.4271/9781468601459

ISBN-Print 978-1-4686-0144-2

To purchase bulk quantities, please contact: SAE Customer Service

E-mail: CustomerService@sae.org
Phone: 877-606-7323 (inside USA and Canada)
724-776-4970 (outside USA)
Fax: 724-776-0790

Visit the SAE International Bookstore at books.sae.org

Chief Product Officer
Frank Menchaca

Publisher
Sherry Dickinson Nigam

Director of Content Management
Kelli Zilko

Production Associate
Erin Mendicino

Manufacturing Associate
Adam Goebel

dedication

I thank my wife and soulmate, Marlene, whose support and encouragement helped me find the time to access all the information necessary to write this book. In addition, this is to my three grandsons: Zachary, Nicholas, and Tyler who are a new generation of scientists and engineers.

contents

preface

Managing Electric Vehicle Power focuses on power management of electric vehicles, which employ high power levels. To ensure proper electric vehicle (EV) operation, this power must be controlled, which is the role of power management. Following are descriptions of the chapters in the book.

Chapter 1 describes the two power-management environments in a typical EV: high voltage and low voltage. The high voltage environment includes the primary power source, either a traction battery or fuel cell. The low voltage environment is produced by a DC-DC converter that accepts the high voltage input from the primary power source and produces low voltage power for electronic control units (ECUs) and sensors.

Chapter 2 describes power management of a battery-powered EV. Here, the high voltage traction battery is the primary power source for the EV's DC-AC inverter that powers its traction motor. This also involves the use of either an on-board or external battery charger.

In Chapter 3, the discussion covers a fuel-cell powered EV. The most common type of fuel cell for EV applications is the polymer electrolyte membrane (PEMFC) fuel cell.

Chapter 4 describes the various power supply topologies that may be employed in an ECU, including linear and switch-mode types. These topologies affect the size, power consumption and efficiency of the power supplies.

Suggestions for possible future AEC-qualified power management technology parts are described in Chapter 5. This includes semiconductors that have not been qualified, according to AEC-Q100 and AEC-Q101, but are available commercially.

EV's employ power semiconductors as described in Chapter 6, including discrete power semiconductors used to provide switching functions throughout the EV. Power semiconductors include silicon power MOSFETs and insulated gate bipolar transistors (IGBTs). Recently, silicon carbide (SiC) and gallium nitride (GaN) have been employed because they offer performance advantages over silicon devices.

Chapter 7 describes the use of lighting for automotive applications. LEDs are used internally as well as externally. LED lighting provides more light output per watt than incandescent (filament) lighting and high intensity lamps.

In Chapter 8, we discuss the use of electric traction motors that drive the vehicle's wheels. Two types of electric traction motors are now employed in most EVs: AC induction and permanent magnet synchronous motors (PMSM).

Chapter 9 covers circuit protection of ECUs that can be vulnerable to electrical hazards. These hazards include: Electromagnetic Interference (EMI), transients, Electrostatic Discharge (ESD), and power malfunctions. Plus, the ECUs must guard against generation of EMI that affects other vehicles or external electronic equipment.

Closely related to increased electronic content and power management is thermal management that is covered in Chapter 10, the electric vehicle's broad electronic content requires effective thermal management of its electronic circuits, particularly semiconductors, whose performance is temperature sensitive.

Power management of ADAS is described in Chapter 11. ADAS is an informal list of over 20 possible functions. Therefore, very efficient power management will be necessary for ADAS functions.

Power management of autonomous vehicles is more complex than a conventional EV as described in Chapter 12. Autonomous EV functions will require more ECUs and their attendant circuit protection issues, such as EMI and ESD. Plus, this will require very efficient power management to minimize the load on the primary power source.

The subject of Chapter 13 is the CAN Bus. The CAN standard was devised to control actuators, receive feedback from sensors, and receive feedback from ECUs.

The Automotive Electronics Council (AEC) sets qualification standards for the components used in the automotive electronics industry (Chapter 14). AEC standards cover: integrated circuits, discrete power semiconductors, LEDs and other optoelectronic devices, multichip modules, and passive components. This chapter also includes a list of Chinese standards for electric vehicles.

CHAPTER 1

EV Power Management

Power management involves all the power consumed in an electric vehicle (EV), so it impacts the vehicle's performance, safety, and driving range. To provide these vehicle characteristics, power management:

- Ensures that the proper power, voltage, and current are applied to each electronic circuit.
- Ensures that there is isolation between low-voltage and high-voltage (HV) circuits.
- Offers power circuit protection against electrical disturbances that can affect internal or external circuits.

A typical EV (Figure 1.1) employs the power management configuration shown in Figure 1.2. The primary power source in a typical EV is either a Li-ion battery or fuel cell. Batteries that provide power for the traction motor are combined in series and parallel. Fuel cells are packaged in a stack of multiple units. The HV battery can be charged off the AC powerline, whereas the fuel cell is "recharged" with hydrogen.

Regardless of whether it is powered by a battery or fuel cell, the primary power source and a DC–AC inverter produce a HV AC for the traction motor. The primary power source usually provides the 200–800 V that is needed to generate enough power for the traction motor [1, 2, 3]. When a battery is the primary power source, the HV environment of the EV is shown in Figure 1.3.

Low-voltage power is intended for electronic control units (ECUs) as shown in Figure 1.2. Deriving this power from the battery or fuel cell would unbalance it, so a DC–DC converter powered by the primary power source (battery or fuel cell) provides the low voltage. The DC–DC converter supplies regulated low-voltage power to ECU modules and associated sensors.

EV modules differ from those in industrial systems that employ circuit boards mounted in a card rack. Instead, EVs employ rugged, enclosed ECU modules that are

FIGURE 1.1 Typical electric vehicle.

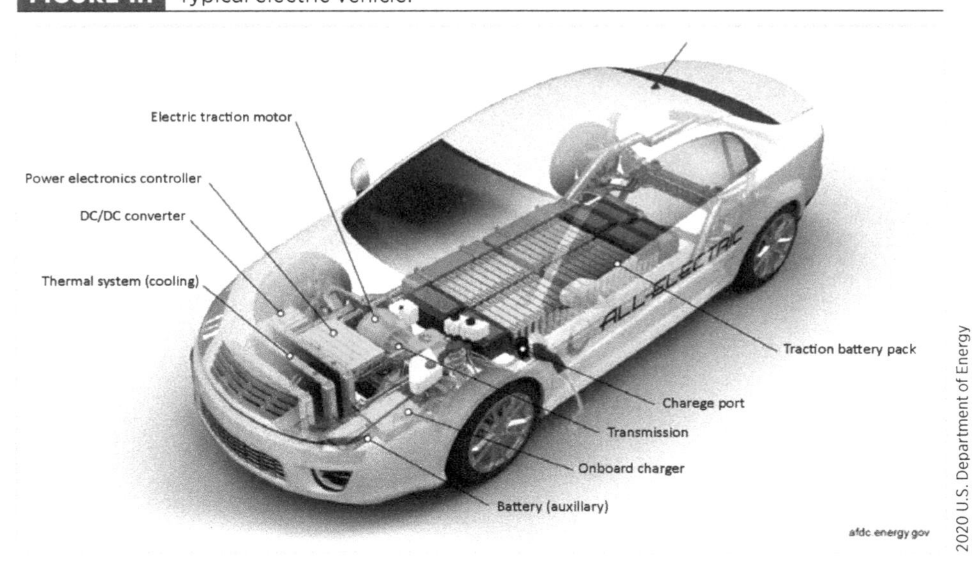

FIGURE 1.2 EV power management with HV battery or fuel cell as the primary power source.

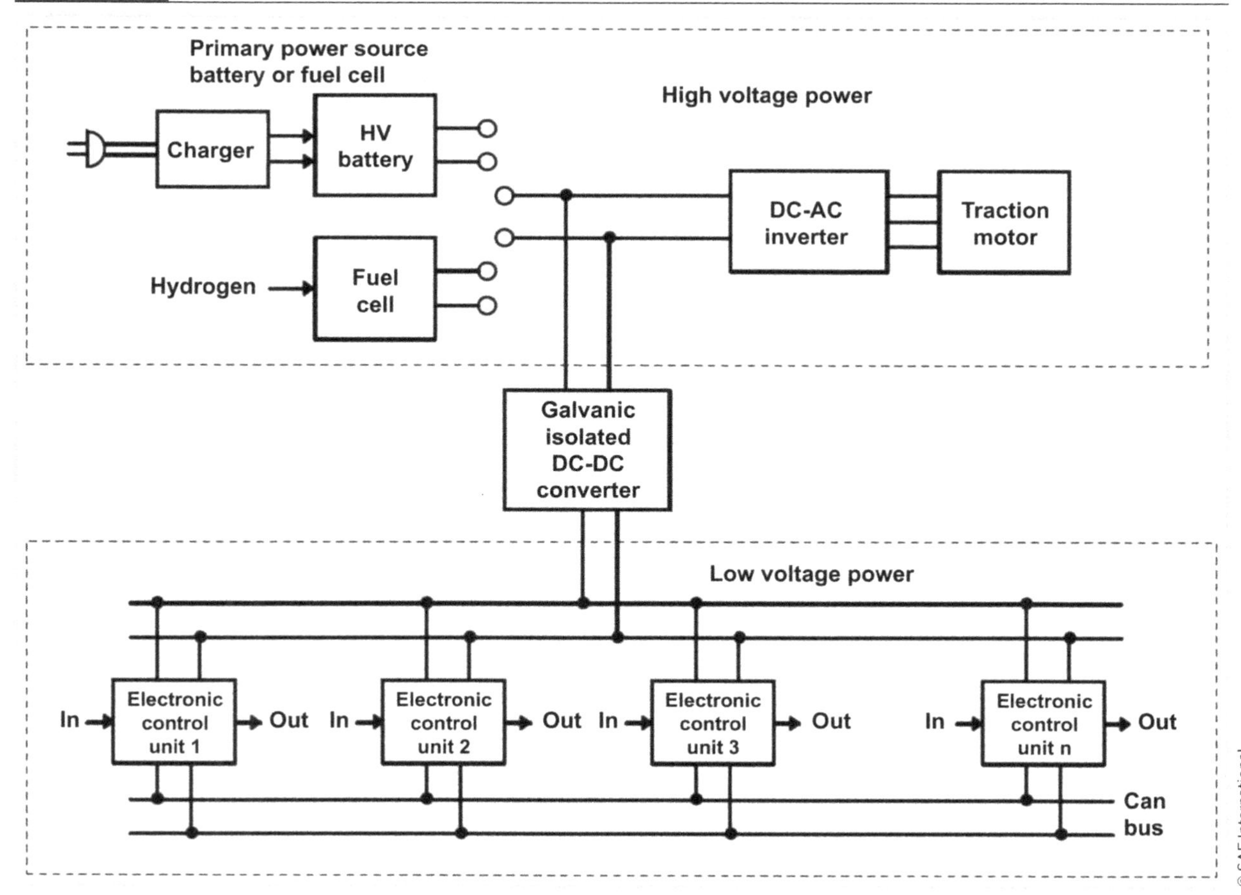

FIGURE 1.3 EV high-voltage power management.

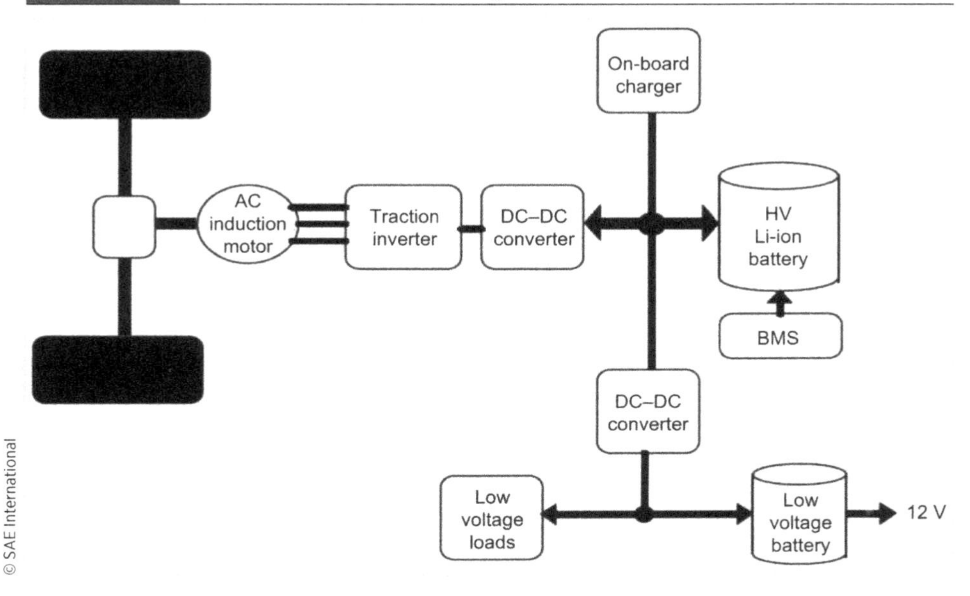

interconnected with cables, whereas industrial systems interconnect the p.c. boards in the card racks.

1.1 DC-AC Inverter

Figure 1.4 shows a traction inverter module circuit that converts the DC from the primary power source to AC for the three-phase AC induction motor that drives the vehicle's propulsion system [4, 5]. It also plays a significant role in capturing energy from regenerative braking and feeding it back to the battery.

Common techniques used to control the DC–AC inverter are pulse-width modulation (PWM) or pulse-frequency modulation (PFM). PWM is the most commonly used

FIGURE 1.4 Traction inverter accepts a high-voltage DC input and produces a three-phase AC output to power the induction motor.

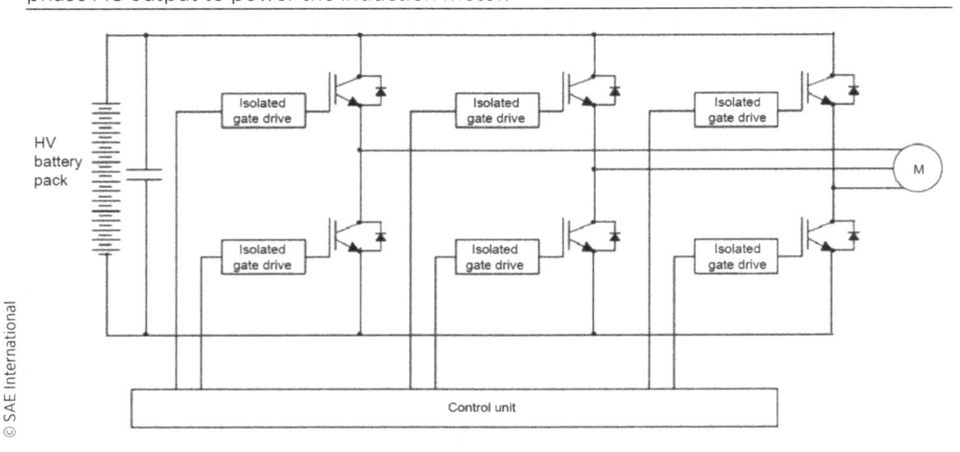

technique. PWM varies the width of pulses required for the switching of transistors in an inverter in order to generate an output waveform composed of many narrow pulses in each cycle. The control unit makes the average voltage of the modulated output pulses sinusoidal.

Separate packaged "box-type" inverters are commonly used by many automotive manufacturers and system suppliers. These inverters using power modules benefit from ease of assembly, while maintaining a modular design approach.

Inverter outputs may provide pure sine waves or modified/quasi sine wave (MSWs). True inverters transform DC into a smoothly varying alternating current very similar to the kind of genuine sine wave. Modified sine wave inverters produce a kind of "rounded-off" square wave that is a much rougher approximation to a sine wave. These inverters consume more power than a pure sine wave so there may be some risk of overheating with MSW inverters.

Another design philosophy applies a more integrated inverter design, where small and fully encapsulated power modules are placed within the mechanical compartment of the drivetrain. Here, the electric machines can be designed as a fully integrated part of the transmission, by integrating the power inverter into the same housing.

The demand for higher levels of integration of power modules into inverters is growing. The concept of an electrical machine including all the power electronics needed for a frequency inverter, matching requirements for cooling, vibration, and robustness, is very attractive when it comes to the sensitive discussion of volume or weight restrictions and overall system cost.

Adequate cooling and ventilation are important in keeping the components operational. For this reason, inverter/converter installations have dedicated cooling systems.

1.2 Galvanic Isolation

Power management for EVs requires some form of isolation using electrical or magnetic separation between the HV and low-voltage circuits [6]. Isolation provides a barrier that dangerous HVs cannot pass in the event of a fault or component failure. This barrier ensures that the electrical circuits are safe from HV damage [7].

Transformers provide isolation by magnetically coupling the primary side AC signal to the secondary winding. Optocouplers provide isolation by transferring a signal using an LED light source that hits a light-sensitive diode in an enclosed case.

In achieving isolation between the control side and the output side of the gate driver in the inverter, the Analog Devices ADuM4137 uses a high-frequency carrier that transmits data across the isolation barrier with *i*Coupler chip-scale transformer coils separated by layers of polyimide isolation [8]. The ADuM4137 uses positive logic on/off keying (OOK) encoding where a high signal is transmitted by the presence of the carrier frequency across the *i*Coupler chip-scale transformer coils. Positive logic encoding ensures that a low signal is seen on the output when the input side of the gate driver is unpowered. A low state is the most common safe state in enhancement mode power devices and can drive in situations where shoot-through conditions are present. The architecture of the ADuM4137 is designed for high common-mode transient immunity and high immunity to electrical noise and magnetic interference. Radiated emissions are minimized with a spread spectrum OOK carrier and differential coil layout. Figure 1.5 shows the internal circuit of the ADuM4137.

FIGURE 1.5 Analog Devices' ADuM4137 4.0 A isolated, single channel gate driver.

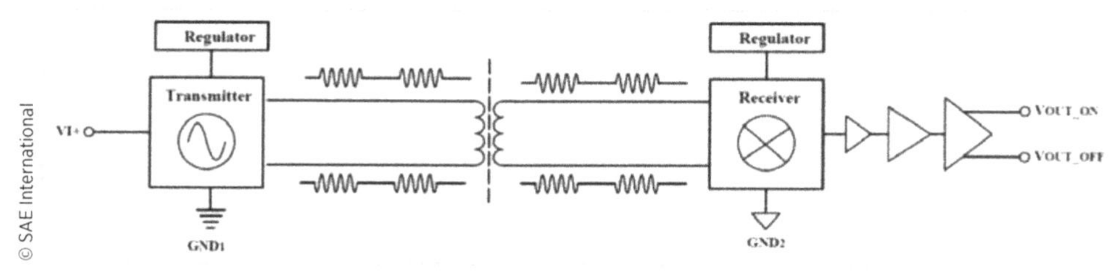

1.3 Primary DC-DC Converter

An example of an automotive primary DC-DC converter is one from LG Innotek (Figure 1.6). Table 1.1 lists its specifications [9].

The BSC624-12 V from BRUSA is a bidirectional DC-DC converter for EV applications featuring galvanic isolation between the HV and the low-voltage circuit (Figure 1.7).

This BSC624-12V employs a series resonant transformer stage. Two buck/boost converters working in a geared mode provide ripple reduction. You can set the desired output voltage in the respective operating mode. The resonant topology of the transformer stage and the auto commutation of the buck/boost converter limits losses and achieves

FIGURE 1.6 LG Innotek DC-DC converter for an EV accepts a high-voltage input from an EV battery stack and produces 9-16 V for ancillary devices.

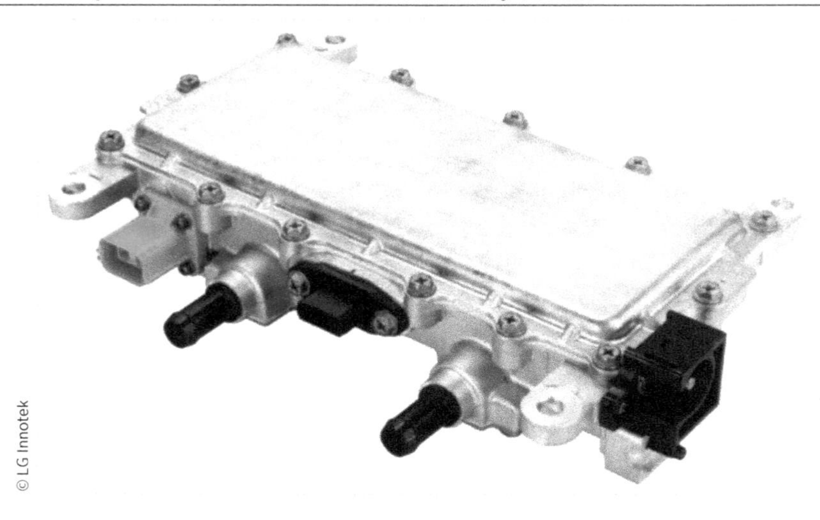

TABLE 1.1 Specifications for LG Innotek DC-DC converter.

Parameter	Description
Capacity (kW)	1.5/2.0/3.0/3.5
Input voltage (V)	200-450
Output voltage (V)	9-16
Efficiency (%)	95 (max.)
Sealing effectiveness	IP6K9K
Cooling method	Water cooling/air cooling

FIGURE 1.7 Brusa's BSC624-12 V converts DC input to AC and uses a transformer to provide galvanic isolation for its DC-output.

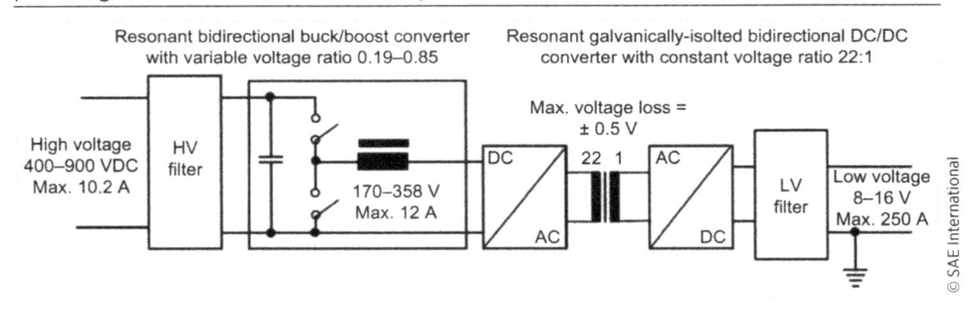

TABLE 1.2 BSC624-12V specifications.

Parameter	Description
Voltage range HV	220–450 V
Nominal voltage LV	14.0 V
Voltage range LV	8–16 V
Continuous current LV	200 A
Max. current LV	250 A
Continuous power buck mode	2.8 kW
Max. power buck mode	3.5 kW
Efficiency	94.4%
Ambient operating temp.	-40°C to +85°C
Cooling system	Liquid-cooled
IP protection	IP65
Weight	4.8 kg

excellent EMC properties. It has a compact, vibration-resistant, lightweight enclosure prioritized to enable its use in almost all applications and installation locations. Table 1.2 lists the specifications of the BSC624-12V [10].

1.4 Electronic Control Unit

An ECU is actually an embedded, computerized electronic system used as the basis for a vehicle's operational network. ECUs read various sensor input data and control many different functions in a vehicle, such as motor control, door control, seat control, speed control, and so on. They include a programmable processor to control their required functions.

A typical ECU consists of a printed circuit board (PCB) inside a metal enclosure. The PCB is mounted inside the enclosure with waterproof cutouts for the input/output connectors. Underneath there is a bottom plate that seals the PCB from the outside and on top there is a cover and collar that provides basic protection for the connectors. There is also a vent for equalizing the pressure inside and grommets for shock and vibration mitigation. The housing is usually made from aluminum, a relatively low-cost, low-weight material with good heat-transfer properties [11].

Depending on where it is mounted, the ECU may be subjected to harsh weather and operating conditions. This includes high vibration levels, high temperature, solar radiation, as well as exposure to grease, oil and water, and road dirt. Therefore, the ECU must be housed in a rugged enclosure to prevent contamination from external forces.

Figure 1.8 is a generic ECU similar to those used in today's automobiles [12]. Its microcontroller includes a nonvolatile memory (NVM) that allows software to be updated in the field. It accepts a voltage input from the main DC–DC converter ($+V_{IN}$) and uses an internal power supply to produce 5 V or 3.3 V for the interface signaling and 1–1.5 V for the processor. The ECU's internal power supply must be electronically robust to survive large voltage fluctuations and transients that may be present on the main DC–DC converter's input. Additionally, the ECU must prevent electrostatic discharge (ESD) and electromagnetic interference (EMI) events that may be present on the main DC–DC converter's input. Also it is required to withstand transients from outside connections, including the CAN bus, sensor, and actuator interfaces. All circuits around the microcontroller form a shield that protects it from voltage excursions and events that might destroy it.

Modern EVs may have up to 80 ECUs. Usually, the main DC–DC converter supplies a single low-output voltage to the ECUs [13]. If other voltages are required, the ECU produces them internally. Each ECU has its own integrated power supply that produces the appropriate power and voltage levels for its processor and other digital circuits. In most cases, the ECU also provides the source voltage for analog sensors [14, 15, 16].

An ECU is actually a generic name for electronic/electrical modules in an EV [17]. Types of ECU modules may include:

- Powertrain control module (PCM)
- Power steering control module (PSCM)
- Transmission control module (TCM)
- Electronic brake control module (EBCM)

FIGURE 1.8 Generic ECU includes a microcontroller and system interfaces.

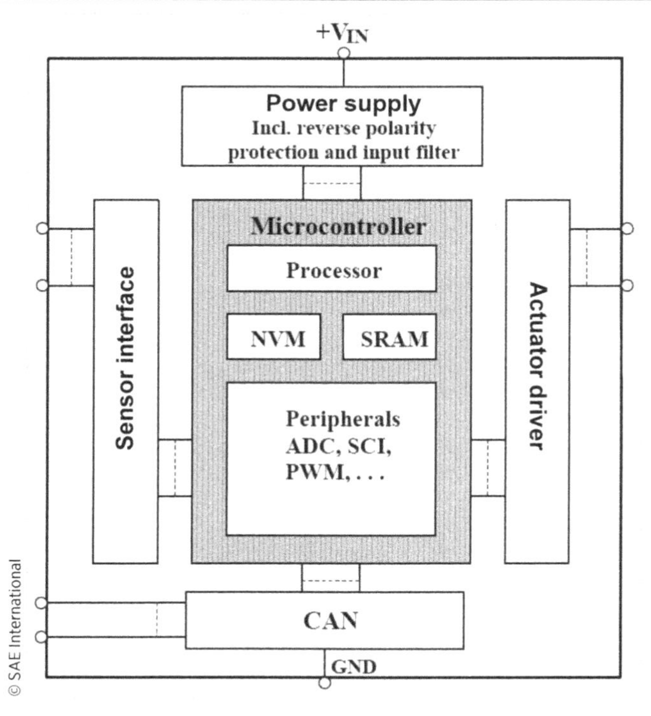

- Central control module (CCM)
- Central timing module (CTM)
- General electronic module (GEM)
- Body control module (BCM)
- Suspension control module (SCM)

There is also a need to manage the ECU power supply according to system performance targets. However, given the noisy operating environment of vehicles with their multiple electronic subsystems, balancing power demands with power constraints can be a challenge. Therefore, precision, flexible, small solution module sizes are essential to meet these stringent system requirements.

Managing the electrical and power considerations of a vehicle's ECU requires a delicate balance. Processors, memories, displays, and other components need well-regulated voltages at various current levels. The regulators, in turn, must be efficient in order to deliver the power needed to run these critical circuits without too much heat dissipation. When there are multiple power rails, there may be many more voltage and current spikes to deal with. In addition, certain voltage rails in a car have specific voltage accuracy requirements.

There is also the ECU's electrical and thermal environment to consider. There can be large and sudden voltage drops when the car is started in various temperature situations, such as cold cranking and warm cranking. Cars may have RF electrical noise from both internal and external sources, causing EMI that can hamper performance in various vehicle subsystems.

Today's EVs have an increasing number of ECUs usually in close proximity in a very confined space. Outside, everything from mobile phones to transmission towers emit noise that can affect the car's performance. Therefore, ECUs must be immune from these noise sources and should not emit EMI that affects other electronic systems.

There is also the possibility of load transients like those that occur when the processor suddenly gets an increased power demand and draws more current. For instance, the processor could be in standby mode at one moment, consuming about one-third of its peak power. Then, when the processor is activated, it could draw the full amount of its current. In this scenario, the power supply's output voltage could temporarily dip and then bounce around before settling at its target voltage. The ability to deal with these load transients requires a well-designed power supply that can manage the output voltage swing and prevent it from impacting system performance.

1.4.1 Centralized vs. Decentralized ECUs [18, 19]

Automotive ECUs control a broad range of functions from motor drives to infotainment systems. Adding a new feature usually requires a separate decentralized (single function) ECU along with its associated sensors, display, and wiring. All these decentralized ECUs and their accessories occupy valuable space, complicate certification testing, and increase the vehicle's cost and weight.

Instead of decentralized (multiple function) ECUs, a single, high-performance, centralized ECU can execute the functions of several systems. Automakers who undertake ECU centralization also recognize the software that accelerates innovation and allows fast response to emerging consumer demands. ECU centralization supports the software-defined EV by making it easier to add and update software-based features. This design approach represents a transition from fixed-function decentralized ECUs to software-defined ECUs that reduce an automakers' development and production costs and greatly increase their agility.

The advantage of centralized ECUs is:

- Addition of advanced driver assistance systems (ADAS) technology can require an increase in computing power for the real-time signal processing of multiple sensors. EVs equipped with more ADAS technology will drive centralization of ECUs compared with equally or less ADAS-equipped EVs.
- The centralized approach could facilitate advanced software development and potentially open up new revenue streams, for example, over-the-air updates.
- Lower design and manufacturing costs because there will be fewer ECUs to design, build, certify, inventory, and assemble.
- Centralization may optimize wiring and sourcing efficiency via increased bundling because this ECU architecture affects vehicle weight and cost.
- Centralized systems have fewer operations that may fail because they require simpler protocols and fewer connections than multiple, decentralized ECUs.
- Centralized ECUs can increase reliability because they require fewer components.
- Centralizing ECUs require fewer development teams and simplified processes, which can lead to shorter development cycles, whereas decentralized ECUs means more teams must collaborate and communicate efficiently.
- Centralized high-power ECUs could lay the foundation for developing fully autonomous driving, thereby setting up EVs to be ready for future mass-market vehicles.
- Faster time-to-market because there are fewer discrete ECUs to test and validate.
- Increased security due to a smaller attack footprint to protect.
- One complication is that centralization may require additional in-house design skills and resources.

1.4.2 ECU Enclosure Protection

The IP Code, International Protection Marking, is from IEC standard 60529 that classifies and rates the degree of protection provided by mechanical castings and electrical enclosures against intrusion, dust, accidental contact, and water. It intends to provide users with detailed numerical coding information about related terms. ECUs are housed in an enclosure, whose "IP Code" defines the levels of their sealing effectiveness against solid objects and liquids.

Table 1.3 lists ECU enclosure ratings that consist of the letters "IP" followed by two digits.

1.4.3 ECU Software

AUTomotive Open System ARchitecture (AUTOSAR) is an industrial alliance of original equipment manufacturers (OEMs), component suppliers, and hardware vendors that established a de facto software standard for electric and electronic architectures of vehicles and their associated ECUs [20]. It focuses on the system and specifies the ways of integrating software components using a predefined and standardized software layer. It provides the advantage that all application software running on an ECU is hardware

TABLE 1.3 IP rating [21].

IP first number—protection against solid objects		
Code	**Effective against**	**Description**
X	—	No data available
0	No special protection	No protection against contact and ingress of objects
1	>50 mm	Any large surface of the body, but no protection against deliberate contact
2	>12.5 mm	Fingers or similar objects
3	>2.5 mm	Tools, thick wires, etc.
4	>1 mm	Most wires, slender screws, large ants, etc.
5	Dust protected	Must not enter in sufficient quantity to interfere with satisfactory equipment operation.
6	Dust tight	No ingress of dust, complete protection against contact (dust tight).
IP second number—protection against liquids		
X	—	No data available
0	None	No protection
1	Dripping water	No harmful effect when mounted in an upright position and rotated at 1 RPM.
2	Dripping water when tilted at 15°	No harmful effect when the enclosure is tilted at 15°.
3	Spraying water	No harmful effect from water falling as a spray at an angle of 60°.
4	Splashing of water	No harmful effect from water splashing against the enclosure from any direction.
5	Water jets	No harmful effect from water projected by a 6.3 mm nozzle against enclosure from any direction.
6	Powerful water jets	No harmful effect from water projected in powerful jets with 12.5 mm nozzle against enclosure from any direction.
6K	Powerful water jets with increased pressure	No harmful effect from water projected in powerful 6.3 mm nozzle jets against enclosure from any direction under elevated pressure.
7	Immersion up to 1 m depth	No harmful effect from ingress of water when the enclosure is immersed in water (up to 1 m) under defined conditions or pressure and time.
8	Immersion 1 m or more depth	No harmful effect from continuous immersion in water under conditions specified by the manufacturer.
9K	Powerful high-temperature water jets	No harmful effect from close-range, high-pressure, high-temperature spray.

independent, through an abstraction layer and an AUTOSAR runtime environment. It describes:

- Basic software modules
- Application interfaces
- Standardized exchange format
- Scalability to different vehicle and platform variants
- Transferability of software
- Safety requirements

You can use the basic layered software modules of AUTOSAR architecture in vehicles of different manufacturers and electronic components of different suppliers. It aids in mastering the growing complexity of automotive, electronic, and software architectures. AUTOSAR is said to pave the way for innovative electronic systems that further improve performance, safety, and environmental friendliness, and to facilitate the exchange and update of software and hardware over the service life of the vehicle.

AUTOSAR complies with ISO 26262, which defines functional safety for automotive equipment applicable throughout the life cycle of all automotive electronic and electrical safety-related systems. Functional safety relates to electrical and electronic systems as well their mechanical subsystems.

ISO 26262 provides:

- Automotive safety life cycle (management, development, production, operation, service, decommissioning) and supports tailoring the necessary activities during these lifecycle phases.
- Functional safety aspects of the entire development process (including such activities as requirements specification, design, implementation, integration, verification, validation, and configuration).
- Automotive-specific risk-based approach for determining risk classes (automotive safety integrity levels, ASILs).
- Validation and confirmation measures to ensure a sufficient and acceptable level of safety is being achieved.

1.5 MCU

Chroma's CP Series motor control unit (MCU) with DSP-based control is specifically intended for EVs (Figure 1.9) [22]. Embedded with an efficient IGBT module for high power density and reliability, its compact and robust design is suitable for electric vehicle e-Drive system integration, including pure battery EV and plug-in hybrid EV.

This MCU configures motor speed and torque after receiving comments from VCU (vehicle control unit) via CAN bus communication. The MCU converts the primary DC power source to AC power supply to drive propulsion motor. During vehicle braking, it

FIGURE 1.9 Chroma CP Series MCU.

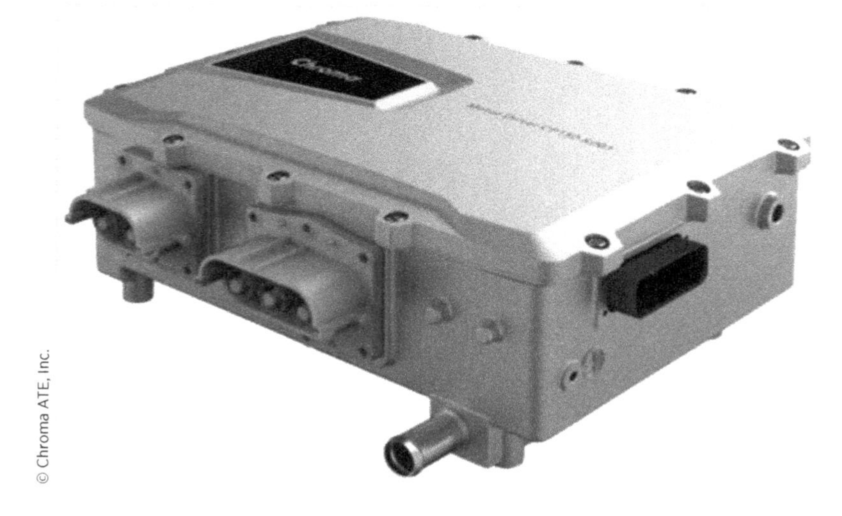

can regenerate DC power back to battery pack for charging. An efficient cooling system enables its high power density and performance. Protection includes against over current, over voltage, and over temperature.

CP Series also comes with integrated dual-mode MCU: one larger power controller for traction motor and one smaller power controller for integrated starter generator (ISG). Main application is PHEV, enabling four (×4) driving modes for the vehicle. ISG controller can start the engine of the range extender, control the ISG, convert AC power to DC which charges the battery pack, or can provide direct DC power to traction motor MCU for lower speed driving. Table 1.4 lists the specifications for the CP105-47 MCU.

An ECU intended as a supervisory controller for electric or hybrid vehicles is the Ecotrons VCU (Figure 1.10). Features include:

- ISO26262 functional safety support.
- Main processor and a monitor processor built-in for safety monitoring.
- Basic software, or BSW, and supports all typical input/output drivers for vehicle controls. The BSW is encapsulated in the Matlab/Simulink environment, and the user can develop the control system with 100% model based design methods.
- Hardware is abstracted from the application software that relieves the controls engineer from the challenge of the microprocessor configuration and embedded real-time software.
- CAN bus-based reprogramming tool with properly configured protocols. It can be used for OTA (over-the-air) programming.
- Support for the CCP/XCP-based CAN bus calibration tools, like INCA, CANape, as well as the cost-effective EcoCAL, developed in-house.

TABLE 1.4 Chroma CP105-47 MCU.

Input	
Nominal battery input	360 VDC
Nominal battery input range	250–450 VDC
Working voltage range	300 VDC
Output	
Continuous delivered power	30 kW (ISG) 53 kW (TM)
Peak delivered power	47 kW (ISG) 105 kW (TM)
Continuous output current	100 Arms (ISG) 200 Arms (TM)
Maximum output current	160 Arms, ≤30 s (ISG) 530 Arms, ≤10 s (TM)
Peak efficiency at 360 VDC	≥95%
Temperature range	
Operating temperature	-40°C to 105°C
Storage temperature	-40°C to 115°C
Others	
Cooling	Liquid, 20 L/min -30°C to 65°C
Communication	CAN Bus 2.0B
Environment	IP67
Weight	24.0 kg

FIGURE 1.10 Ecotrons vehicle control unit (VCU) for electric vehicles.

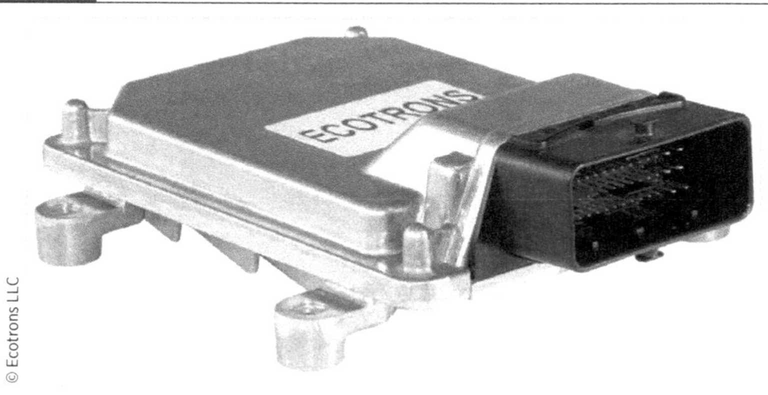

© Ecotrons LLC

Normally, the VCU is intended as a centralized controller for an electrical vehicle. VCU can do other jobs like gateway, body control, but still it is used as vehicle control unit. Basically it can control everything in EV (directly or indirectly), it can control the motor by controlling the MCU, it can control window, it can control OBC, and it needs to talk with BMS.

Table 1.5 lists the specifications for the Ecotrons EV2274A [23].

1.6 ASIL

Automotive Safety Integrity Level (ASIL) describes the risk classification of a vehicle's functions as described in the the ISO 26262 standard: *Functional Safety for Road Vehicles*. This standard defines functional safety as the "absence of unreasonable risk due to hazards caused by the malfunctioning behavior of electrical/electronic systems." You determine ASIL by performing a risk analysis of the specific function by investigating its impact on the severity, exposure and controllability of the vehicle.

There are four ASIL categories: ASIL A, ASIL B, ASIL C, and ASIL D, where ASIL D specifies the highest integrity requirements and ASIL A specifies the lowest. In addition, there is the QM category that does not dictate any safety requirements.

ASIL D functions require highest level of safety assurance against hazards, so they include airbags, antilock brakes, and power steering. In contrast, rear lights require only an ASIL A grade. ASIL B covers headlights and brake lights, and cruise control is usually ASIL C.

1.7 Testing EVs [24]

The main HV traction battery or fuel cell is a key item that must be tested to ensure a safe and efficient EV. Additionally, the manufacturer should be aware of the characteristics of all components that could affect long-term, power source performance and compatibility with charging equipment. If these components are not evaluated properly, they could cause power failures and danger to vehicle operators in hazardous conditions [25].

The EV manufacturer can purchase specific instruments to perform the testing or have a test organization provide the testing, evaluation, and certification to regional and

TABLE 1.5 Ecotrons EV 2274A vehicle control unit (VCU).

Parameter	Description
Main processor	NXP MPC57xx, 150–200 MHz 2.5–4 MB Flash Memory 192–384 kB SRAM Floating Point Capable
Monitor processor	(SBC) MC33CFS8500
Supply voltage	9–32 VDC
Peak voltage	36 VDC
Reprogramming	Bootloader, CCP protocols
CAN bus	3 Channels, 1 Channel Specific Frame CAN Wake Up, 1 Channel Ordinary CAN Wake Up
LIN bus	1 Channel
EEPROM	64K
Sensor supply output	5 Channels 5 V
Analog input	15 Channels, 12-bit Supports both 0 and 5 V 0–36 V inputs
Digital input	17 Channels7 low effective, 10 high effective
Frequency signal input	4 Channels
Low side drivers	11 Channels at 0.5 A 5 Channels at 1 A 2 Channels at 3 A 4 Channels can be configured as PWM output
High side drivers	3 Channels at 0.5 A5 Channels at 1 A 2 Channels at 3 A 2 Channels can be configured as PWM output
Operating temperature	−40°C–85°C
Working humidity	Satisfying 0%–95%, noncondensing
Protection category	IP67
Pin numbers	121
Dimensions	207 × 150 × 36 mm
Housing material	Aluminum
Weight	≤600 g
Mechanical characteristics	Vibration, shock, drop test as in ISO16750

international EV standards, according to those listed in Table 1.6. Some tests may require replication of environmental conditions, such as temperature extremes, salt, ozone, vibration, shock, and impact, to simulate hazards that may arise during vehicle transportation or operation.

Testing of traction batteries in EVs requires checking the charge and discharge functions for different values of current, voltage, power, or resistance. The test system should provide a precision integrated solution for high-power battery pack tests. Accurate sources and measurements ensure the test quality that is suitable for performing exact and reliable testing that is crucial for battery pack incoming or outgoing inspections, as well as capacity, performance, production, and qualification testing [26].

If ADAS is employed in the vehicle, the power applied to each of the associated electronic systems should be checked. Also, it may be necessary to sequentially apply and remove power from these systems. In addition, the vehicle's HV traction battery and associated DC–DC converters should be checked to make sure they can handle the power requirements of the ADAS functions.

TABLE 1.6 SAE Standards related to the design and testing of EVs.

CISPR 25	Vehicles, boats and internal combustion engines - Radio disturbance characteristics - Limits and methods of measurement for the protection of onboard receivers
IEC 61000-4-4	Electromagnetic compatibility (EMC) - Testing and measurement techniques - Electrical fast transient/burst immunity test
IEC 61000-4-5	EMC - Testing and measurement techniques - Surge immunity test
IEC/EN 62660-1	Specifies performance and life testing of secondary lithium-ion cells used for propulsion of electric vehicles including battery electric vehicles (BEV) and hybrid electric vehicles (HEV).
IEC/EN 62660-2	Standard test procedures for the reliability and abuse behavior of secondary Li-ion cells and cell blocks used for propulsion of EVs including battery electric vehicles and HEVs.
IEC 61982	Performance and endurance tests for secondary batteries used for vehicle propulsion applications. Applicable to lead-acid, Ni/Cd, Ni/MH, and sodium-based batteries used in EVs.
IEC 61851-1	EV equipment for charging electric road vehicles, with a rated supply voltage up to 1,000 V AC or up to 1,500 V DC and a rated output voltage up to 1,000 V AC or up to 1,500 V DC.
IEC 62196	Plugs, socket outlets, vehicle connectors, vehicle inlets - Conductive charging of EVs.
ISO 7637-2	Test methods and procedures to ensure the compatibility to conducted electrical transients of equipment installed on vehicles fitted with 12 V or 24 V electrical systems.
ISO 7637-3	Defines immunity testing for "transient transmission by capacitive and inductive coupling via lines other than supply lines."
ISO 10605	Specifies the electrostatic discharge (ESD) test methods necessary to evaluate electronic modules intended for vehicle use.
ISO 11452-2	Specifies an absorber-lined shielded enclosure method for testing the immunity (off-vehicle radiation source) of electronic components for passenger cars and commercial vehicles regardless of the propulsion system.
ISO 11452-2	Road vehicles - Component test methods for electrical disturbances from narrowband radiated electromagnetic energy - Absorber-lined shielded enclosure.
ISO 11452-4	Road vehicles - Component test methods for electrical disturbances from narrowband radiated electromagnetic energy: Harness excitation methods
ISO 23273	Essential requirements for fuel cell vehicles (FCV) with respect to the protection of persons and the environment inside and outside the vehicle against hydrogen-related hazards.
ISO 6469	Safety specifications for electric road vehicles addresses onboard rechargeable energy storage systems for the protection of people inside and outside the vehicle as well as safety means and protection against electrical failures.
FMVSS 305	Federal Motor Vehicle Safety Standards specify design, construction, performance, and durability requirements for motor vehicles and regulated automobile safety-related components, systems, and design features.
UL2251	Requirements cover plugs, receptacles, vehicle inlets, and connectors rated up to 800 A and up to 600 V AC or DC, for conductive connection systems used with electric vehicles in accordance with National Electrical Code (NEC), ANSI/NFPA-70.
SAE J2344	Identifies and defines the preferred technical guidelines relating to safety for vehicles that contain high voltage, such as EVs, HEVs, PHEVs, fuel cell vehicles (FCV), and plug-in fuel cell vehicles (PFCV) during normal operation and charging, as applicable.
SAE J2578	Identifies and defines the preferred technical guidelines relating to the safe integration of fuel cell system, the hydrogen fuel storage and handling systems as defined and specified in SAE J2579, and electrical systems into the overall fuel cell vehicle.
SAE J2464	Test procedures covering vehicle applications and electrical energy storage devices, including individual RESS cells (batteries or capacitors), modules, and packs.
SAE J2929	Defines acceptable safety criteria for a lithium-based rechargeable battery system for use in a vehicle propulsion application as an energy storage system connected to a HV power train.
SAE J1772	Electric vehicle conductive charge coupler covers the physical, electrical, communication protocol, and performance requirements for an EV conductive charge system and coupler.
SAE J2380	Describes the vibration durability testing of a single battery (test unit) consisting of either an electric vehicle battery module or an electric vehicle battery pack.

1.8 System Tests

The power conversion section of an EV consists of several types of power electronic units, including the AC or DC EVSE (EV Supply Equipment), onboard charger, DC/DC converter, motor driver, and so on. Particularly, regardless of whether the vehicle uses an onboard or external charger, it should be checked to ensure that it is capable of charging the battery within the allotted time [27, 28].

The EV test system must address the specialized requirements involved in testing the power electronics during the development phase, as well as the production phase. Advanced features, such as automatic test data recording of statistical analytical reports, should be available for design reviews or product improvement.

To provide optimum vehicle operating conditions, maintain serviceability, and minimize the possibility of a mechanical exposure to electric shock, the test equipment should provide dynamic diagnostic capabilities. This includes voltage/current parameter measurement reading, CAN bus interface, diagnostic reporting, and so on.

Extended reporting capabilities should be available to allow testers to integrate different types of presentations. Users may also edit and store report formats for future use, thus saving time creating test reports. Test conditions defined in the test program as well as the test readings can be stored and analyzed. The report and raw data may be printed out or stored in a file.

An electrical safety test is important for EVs because most drivers use their cars every day. Electrical safety covers the power system, the charging system, the power wiring, the charging line, the charging connector, the charging station, and so on.

The demand for charging stations is steadily growing as EVs gain popularity. The EV charging station has implemented sophisticated electronics for metering, controlling, and measuring the amount of energy required and transferred to the vehicle. In order to keep the charging station in optimum operating condition and maintain its accuracy, frequent service and calibration are required.

An EV's power electronic units include:

- AC or DC EVSE (EV Supply Equipment)
- Onboard charger
- DC/DC converter
- Motor driver

The Chroma 8000 ATS (Automatic Test System) addresses the specialized requirements involved in testing these power electronics units during the development and production phases (Figure 1.11). This ATS has an open architecture, allowing the user to easily integrate various instruments. It includes a wide range of hardware choices such as AC/DC power supplies, electronic loads, power analyzers, oscilloscopes, digital multimeters, as well as various digital/analog I/O cards. This flexibility combined with an open architecture gives the test engineer an adaptable, powerful, and cost-effective test system for the EV/HEV power electronics.

This test system includes a sophisticated test executive, which includes pre-written test items. Users may also create new test items by using the test item editor function. This provides the flexibility to expand your test library without limits. The Chroma 8000 ATS' ability to satisfy the test requirements for multiple power electronic units is key to keeping consistency and reducing costs during the transition between R&D and production. This ATS also has analytical reports for later design reviews or product improvements.

FIGURE 1.11 Chroma 8000 ATS for EVs.

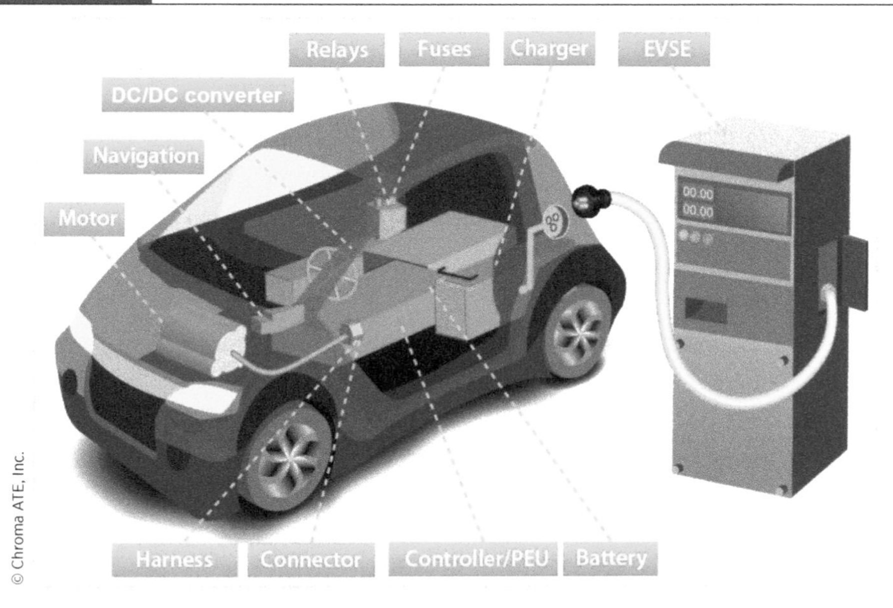

References

1. Sridhar, N., "Driving the Future HEV/EV with High Voltage Solutions (Rev. B)," Texas Instruments SLYY052B-6, May 16, 2018.
2. Frenzel, L., "High Voltage Solutions Take over HEV/EV Designs," Electronic Design, July 2, 2018.
3. Flah, A., Lassad, S., and Mahmoudi, C., "Overview of Electric Vehicle Concept and Power Management Strategies, 2014 Conference on Electrical Sciences and Technologies in Maghreb (CISTEM 2014), November 1, 2014, doi: 10.1109/CISTEM.2014.7077026.
4. Gable, C. and Gable, S., "Inverters and Converters in Hybrids and EVs (Electric Vehicles)," ThoughtCo, January 12, 2019.
5. Taranovich, S., "The Latest in Electric Vehicle Power Management," EDN, October 17, 2016.
6. Littelfuse, "Circuit Protection on High Reliability Electric Vehicles (EVs)," Littlefuse, March 14, 2017.
7. Mersen, "Overcurrent Protection Guide for DC Battery Applications," Mersen, July 24, 2018.
8. Analog Devices, "ADuM4137, High Voltage, Isolated IGBT Gate Driver with Fault Detection," Analog Devices, August 7, 2019.
9. LG Innotek, "DC-DC Converter For Automotive," LG Innotek.
10. BRUSA, "BRUSA BSC624-12V DC/DC Converter," BRUSA.
11. Chellswamy, C. and Ramesh, R., "An Intelligent Energy Management and Control System for Electric Vehicle," IEEE Conference on Advanced Communications Control and Computing, Ramanathapuram, India, May 2014.
12. van Dijk, L., "Future Electric Vehicle Networks and ECUs," NXP Semiconductors.

13. Ciulla, V., "Electronic Control Unit," Livabout.com, April 30, 2018.
14. Wikipedia, "Electronic Control Unit," Wikipedia.
15. Embitel, "ECU Is a Three-Letter Word for All the Innovative Features in Your Car, Embitel,
16. Keihin, Electronic Control Units, Keihin, 2019.
17. Mahadevan, A., "What is the Electronic Control Unit in Electric Vehicles," Quora, June 3, 2016.
18. Chatelain, A., Erriquez, M., Mouliere, P.Y., and Schafer, P., "What a Teardown of the Latest Electric Vehicles Reveals about the Future of Mass-Market EVs," McKinsey, May 2018.
19. Intel, "ECU Consolidation Reduces Vehicle Cost, Weight, Testing," Intel, November 15, 2018.
20. Mihai, S., "AUTOSAR-Compliant ECU Design for Electric Vehicles," Fortech, June 12, 2018.
21. Engineering Toolbox, "IP - Ingress Protection Rating," Engineering Toolbox, 2003.
22. Chroma, "CP Series MCU," Chroma.
23. Ecotrons, "Vehicle Control Unit," Ecotrons.
24. National Instruments, "Selecting an Approach to Build Flexible, Cost-Effective ECU Production Test Systems," August 29, 2019.
25. Hawes, M., "Emerging Solutions to Hybrid & Electric Vehicle DC:DC Converter Design and Test," Keysight Technologies, December 4, 2017.
26. Zhang, R., Cheng, X., and Yang, L., Flexible Energy Management Protocol for Cooperative EV-to-EV Charging, IEEE Transactions on Intelligent Transportation Systems, February 6, 2017, 172-174.
27. Sariri, S., Schwarzer, V., Gorbani, R., "Electric Vehicle Interaction at the Electrical Circuit Level," Hawaii National Energy Institute FSEC-CR-2077-18, January 2018.
28. Western Automation, "Electrical Safety for Electric Vehicle Charging," Western Automation, April 28, 2016.

CHAPTER 2

EV Batteries

A battery is one type of primary power source for an EV, the other is a fuel cell. Today, most EV battery packs consist of various versions of lithium-ion batteries, until something better comes along [1–3]. Two types of Li-ion battery packs may be used in an EV, including:

- Lithium-ion batteries deliver 3.7 V, about three times the power of nickel-metal hydride batteries, yet they are light and deliver substantial power.
- Lithium polymer batteries are a rechargeable battery of lithium-ion technology using a polymer electrolyte instead of a liquid electrolyte. These batteries provide higher specific energy than other lithium battery types.

2.1 Typical EV Battery Pack Characteristics

- Most EV batteries consist of various versions of lithium-ion batteries [1–3].
- Large stacks of cells are grouped into smaller stacks that are enclosed modules.
- If a cell in one module fails, you only need to replace its effected module. Full replacement of all the batteries would be more difficult than replacing a single module.
- In the middle, or at the end of the battery cell stack, is a fuse that limits pack current if there is a short circuit.

- A "service plug" or "service disconnect" can be removed to split the battery stack into two electrically isolated halves. Plug removal protects service technicians from the battery's high voltage.
- Relays, or contactors, control the distribution of the battery pack power. Usually, there are at least two main relays that connect the battery stack's two output terminals that supply high voltage to the traction motor.
- Some packs include alternate current paths for pre-charging the drive system through a pre-charge resistor or for powering auxiliary busses that will also have their own associated normally open control relays.
- There are several temperature, voltage, and current sensors. Temperature sensors are located at several locations inside the pack.
- One sensor measures pack current that can be used to track its actual state of charge (SoC).
- A main voltage sensor monitors the voltage of the entire stack.

An MIT team's *Guide to Understanding Battery Specifications* lists battery characteristics, as shown in Table 2.1 [4].

2.2 Battery Management

EVs employ an intelligent battery management system (BMS) that monitors battery operation. The BMS works in real time to monitor the performance of individual battery cells to ensure they are operating properly (Figure 2.1). A typical BMS monitors [5, 6]:

- Voltage: total voltage, voltages of individual cells, minimum and maximum cell voltage, or voltage of periodic taps
- Temperature: average temperature, coolant intake temperature, coolant output temperature, or temperatures of individual cells
- State of charge (SOC) or depth of discharge (DOD): to indicate the charge level of the battery
- State of health (SOH): measurement of the remaining capacity of the battery as % of the original capacity
- State of power (SOP): amount of power available for a defined time interval given the current power usage, temperature, and other conditions
- Coolant flow: for air- or fluid-cooled batteries
- Current: current in or out of the battery

Checking individual battery cells is a critical concern for EV batteries. Among the safety concerns are thermal runaways that could cause a fire in the vehicle's battery. Thermal runaways can be caused by several malfunctions, such as overcharging. To help avoid unsafe events, the BMS must be able to constantly monitor and detect changing operating parameters and take protective action like shutting down a battery cell that is overheating.

Another safety concern is the ability to verify that alarms/alerts are genuine and not a failure in the BMS. Plus, the BMS must have built-in protection functionality that can instantaneously take the proper and most effective action to head off a runaway condition before it potentially becomes unsafe.

TABLE 2.1 Basic battery characteristics.

Parameter	Description
C-rate	A measure of the rate at which a battery is discharged relative to its maximum capacity. A 1C rate means that the discharge current will discharge the entire battery in 1 h. For a battery with a capacity of 100 Amp-h, this equates to a discharge current of 100 A. A 5C rate for this battery would be 500 A, and a C/2 rate would be 50 A.
E-rate	Describes the discharge power. 1E rate is the discharge power to discharge an entire battery in 1 h.
Primary cells	Batteries that cannot be recharged.
Secondary cells	Batteries that are rechargeable.
State of charge (SoC, %)	Present battery capacity as a percentage of maximum capacity. SoC is generally calculated using current integration to determine the change in battery capacity over time.
Depth of discharge (DOD) (%)	Percentage of battery capacity that has been discharged, expressed as a percentage of maximum capacity. A discharge to at least 80% DOD is referred to as a deep discharge.
Internal resistance	Resistance within the battery, generally different for charging and discharging, also dependent on the battery state of charge. As internal resistance increases, the battery efficiency decreases, and thermal stability is reduced as more of the charging energy is converted into heat.
Maximum continuous discharge current	Maximum current at which the battery can be discharged continuously. Usually defined by the battery manufacturer in order to prevent excessive discharge rates that would damage the battery or reduce its capacity.
Maximum 30-s discharge pulse current	Maximum current at which the battery can be discharged for pulses of up to 30 s. Usually defined by the battery manufacturer in order to prevent excessive discharge rates that would damage the battery or reduce its capacity. Along with the peak power of the electric motor, this defines the acceleration performance (0–60 mph time) of the vehicle.
Charge voltage	Voltage that the battery is charged to when charged to full capacity. Charging schemes generally consist of a constant current charging until the battery voltage reaching the charge voltage, then constant voltage charging, allowing the charge current to taper until it is very small.
Float voltage	The voltage at which the battery is maintained after being charged to 100% SoC to maintain that capacity by compensating for the battery's self-discharge.
(Recommended) Charge current	Ideal current at which the battery is initially charged (to roughly 70% SoC) under constant charging scheme before transitioning into constant voltage charging.
Temperature-dependence	Do not charge batteries when their operating temperature is below freezing. Some battery packs have a heating blanket to warm the battery during cold temperature charging. Also, there should be a lower charge current when the battery is cold. Fast charging when the batteries are cold promotes dendrite growth in Li-ion that can compromise battery safety.
Capacity or nominal capacity (Ah for a specific C-rate)	Coulometric capacity, total amp-hours available when the battery is discharged at a certain discharge current (specified as a C-rate) from 100% state-of-charge to the cut-off voltage. Capacity is calculated by multiplying the discharge current (in amperes) by the discharge time (in hours). Capacity decreases with increasing C-rate.
Energy or nominal energy (Wh for a specific C-rate)	"Energy capacity" of the battery, the total watt-hours available when the battery is discharged at a certain discharge current (specified as a C-rate) from 100% state-of-charge to the cut-off voltage. Energy is calculated by multiplying the discharge power (in watts) by the discharge time (in hours). Energy decreases with increasing C-rate.
Cycle life (number for a specific DOD)	Number of discharge-charge cycles the battery can experience before it fails to meet specific performance criteria. Cycle life is estimated for specific charge and discharge conditions. The actual operating life of the battery is affected by the rate and depth of cycles and by other conditions such as temperature and humidity; the higher the DOD, the lower the cycle life.
Specific energy (Wh/kg)	Nominal battery energy per unit mass. Sometimes referred to as the gravimetric energy density. Specific energy is a characteristic of the battery chemistry and packaging. Along with the energy consumption of the vehicle, it determines the battery weight required to achieve a given electric range.
Specific power (W/kg)	Maximum available power per unit mass. Specific power is a characteristic of the battery chemistry and packaging that determines the battery weight required to achieve a given performance target.
Energy density (Wh/L)	Nominal battery energy per unit volume, sometimes referred to as the volumetric energy density. Specific energy is a characteristic of the battery chemistry and packaging. Along with the energy consumption of the vehicle, it determines the battery size required to achieve a given electric range.
Power density (W/L)	Maximum available power per unit volume. Specific power is a characteristic of the battery chemistry and packaging. It determines the battery size required to achieve a given performance target.

FIGURE 2.1 EV battery management system.

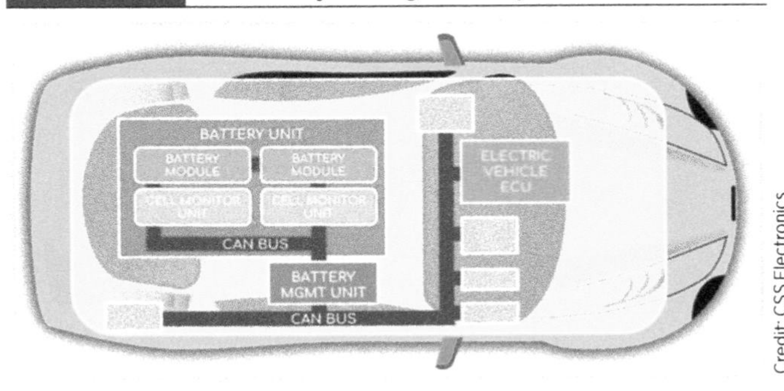

Credit: CSS Electronics

Besides managing battery operation, the BMS often handles other vehicle functions, including the vehicle's desired operating mode, whether it is accelerating, braking, idling, or stopped, which can then activate appropriate power management functions [5, 7].

The health of each individual battery cell in the stack is determined based on its SoC, which measures the ratio of its remaining charge to its cell capacity. SoC uses battery measurements such as voltage, integrated charge and discharge currents, and temperature to determine the charge remaining in the battery.

In EV applications, the SOC is used to determine the vehicle's driving range. It should be an absolute value based on battery capacity when new, not a percentage of current capacity which could result in an error of 20% or more due to battery ageing.

2.3 **Cell Balancing**

Hundreds of battery cells are connected in series and parallel to achieve the total voltage and current requirements to power a DC–AC inverter that drives the traction motor [9]. A major function of a BMS is to balance hundreds of battery cells that provide enough voltage to power the associated traction motors that may operate from 200 to 800 V and dissipate up to 100 kW or more [9]. The battery stack is limited in performance by its lowest capacity cell; once the weakest cell is depleted, the entire stack is effectively depleted.

Parallel cells will balance each other with mutually applied voltage. For cells in series, the weakest cell determines the empty point for the battery pack. The lowest capacity cell will have the lowest voltage and cause end of discharge conditions in battery gauges and under voltage conditions to trip safety circuits. Thus, an undercharged series cell will reduce the entire pack's lifetime.

To keep cells balanced, the individual voltages must be monitored. When a voltage difference between cells becomes too large, a circuit can be enabled to draw more current from the higher cells. Cell balancing techniques can be active or passive.

Passive balancing allows the stack to look like every cell has the same capacity as the weakest cell. Using a relatively low current, it drains a small amount of energy from high SoC cells during the charging cycle so that all cells charge to their maximum SoC. This is accomplished by using a switch and bleed resistor in parallel with each battery cell. The high SoC cell is bled off (power is dissipated in the resistor) so that charging can continue until all cells are fully charged. Passive balancing can also correct for long-term mismatch in self-discharge current from cell-to-cell.

Although passive balancing allows all batteries to have the same SoC, it does NOT improve the run time of a battery-powered system. It provides a fairly low-cost method for balancing the cells, but it wastes energy in the process due to the discharge resistor.

Active balancing employs a DC–DC converter to balance HV stacks of batteries. The high efficiency of a switching regulator significantly increases the achievable balancing current, while reducing heat generation. Active balancing also allows for capacity recovery in stacks of mismatched batteries, a feat unattainable with passive balance systems. In a typical system, more than 99% of the total battery capacity can be recovered.

HV batteries require careful cell matching, especially when drawing heavy loads or operating at cold temperatures. With multiple cells connected in a series string, the possibility of one cell failing is real and would cause a failure. To prevent this from happening, a solid-state switch in some large packs bypasses the failing cell to allow continued current flow, albeit at a lower string voltage. Cell matching is a challenge when replacing a faulty cell in an aging pack. A new cell could have a higher capacity than the others, causing an imbalance.

Initially, a battery stack may have fairly matched cells. But over time, the cell matching degrades due to charge/discharge cycles, elevated temperature, and general aging. A weak battery cell will charge and discharge faster than stronger or higher capacity cells, and thus it becomes the limiting factor in the run time of a system.

2.4 Battery Monitor IC

There are AEC-qualified battery monitor, protector, and cell balancer ICs intended for automotive applications. Texas Instruments BQ76PL455A-Q1 16-Cell Battery Monitor is a battery monitoring and protection IC that is AEC-Q100 qualified. It monitors and detects several different fault conditions, including overvoltage, undervoltage, over temperature, and communication [10]. Six GPIO ports, as well as eight analog AUX ADC inputs, are included for additional monitoring and programmable functionality. A secondary thermal shutdown is included for further protection. Its integrated high-speed, differential, capacitor-isolated communications interface allows up to 16 BQ76PL455A-Q1 devices to communicate with a host.

2.5 Commercial BMS

There are also commercial BMS available, such as the one from Continental (Figure 2.2) that manages all functions of the HV Li-ion battery applied to hybrid and electric vehicles.

Characteristics of the Continental BMS:

- Monitors Li-ion cells for 400 and 800 V systems
- Integrated HV system isolation measurement
- Scalable monitoring of cell temperatures
- Scalable contactor control and monitoring
- Multiple options for battery current measurement
- Integrated cell balancing
- AutoSAR-compliant software

Features

- Calculates SOC and State of Health (SOH) of battery cells
- Controls internal and external actuators like contactors, cooling pumps, etc.
- Scalable design, compact size and weight
- Modular structure including the battery management controller and the cell supervising circuit

FIGURE 2.2 Continental battery management system.

2.6 Battery Safety Issues

NHTSA (National Highway and Transportation Safety Administration), in an October 2017 report "Lithium-ion battery Safety Issues for Electric and Plug-in Hybrid Vehicles," noted that R&D is in progress to achieve greater Li-ion battery performance at lighter weight and lower cost [11]. The report concluded that "the propensity and severity of fires and explosions from the accidental ignition of flammable electrolytic solvents used in Li-ion battery systems are anticipated to be somewhat comparable to or perhaps slightly less than those for gasoline or diesel vehicular fuels. The overall consequences for Li-ion batteries are expected to be less because of the much smaller amounts of flammable solvent released and burning in a catastrophic failure situation."

Another safety concern described in the report was the isolation of HV components to protect passengers and first responders in the event of a crash. The loss of HV isolation can manifest two hazards including short circuit of the battery causing a thermal event or HV potential exposure to humans.

The report pointed out that there are numerous external events or processes in the life of a vehicle that could contribute to damage and failure of a Li-ion battery. In general, the technical literature indicates that, while there are many contributing factors, the primary parameters controlling Li-ion cell and battery performance are temperature and operating voltage.

For each battery chemistry, design, and expected duty cycle, there is a range of temperatures and range of operating voltage in which electrochemistry is dominated by intercalation mechanisms. Outside this range, undesirable side reactions may occur that can lead to self-heating (exothermic reactions) and/or internal electrical shorts (excessive flow of electrons).

Intercalation is a reversible process where a guest molecule or ion is inserted into a host lattice. Intercalation in Li-ion batteries occurs only during the charging and discharging process. A Li-ion battery consists of a positive electrode, negative electrode, and electrolytes. During discharging, the positive Lithium ion moves from the negative electrode and enters the positive electrode through the electrolyte solution. During charging, the opposite of this process occurs.

Exothermic reactions and/or internal electrical shorts may be triggered by manufacturing defects, or mechanical, electrical, or thermal errors, misuse or abuse. If allowed to continue, these reactions or shorts can create conditions for self-heating within the cell, which grow to become uncontrolled increases in temperature and pressure (thermal runaway), and potentially end in venting or catastrophic failure of the cell.

The NHTSA report identified seven primary categories of external causes contributing to failure of Li-ion battery cells:

1. External electrical causes such as external electrical short, overcharging, or overdischarging.
2. External thermal causes, such as exposure to high temperatures or charging at cold temperatures.
3. External mechanical causes, which include excessive shock, impact, compression (crush), or penetration.
4. External chemical contamination, including packaging penetration by corrosive and aggressive agents and contamination of internal components by water, saltwater, or corrosive agents.
5. Service-induced stress and aging causes, such as excess cycling that lead to electrochemical component breakdown, fracture, and crack growth.

6. Cumulative abuse and service causes in which combinations of electrical, mechanical, and thermal abuse (summarized above) and normal charge/discharge duty cycles cause damage to initiate and grow to the point of failure.
7. Errors in design, manufacturing, operation, and maintenance, which induce electrical, mechanical, and thermal abuse causes.

2.7 Venting EV Battery Enclosures

Individual cells in a lithium-ion battery pack need protection, so they are housed in an enclosed module. The battery enclosure protects battery cells from damage while preventing the ingress of water and dirt, which is difficult for an enclosure located on the bottom of a vehicle, where it is constantly exposed to environmental challenges and automotive fluids. Welded battery module construction adds to the complexity of the repair, and this is why battery packs are commonly replaced as a unit [12].

Although the enclosures protect the batteries themselves, lithium-ion batteries produce heat while charging and discharging. Heated air expands inside the battery enclosure, causing pressure to build-up. This pressure can exert excessive force on the battery's seals and deform its housing structure. Therefore, the enclosure must be vented to allow air to travel out of and into the enclosure, while preventing water and other contaminants from entering and damaging the internal components in the enclosure.

Donaldson Company, Inc. has developed an active vent that provides battery enclosure protection. The company's protective vents rely on a microporous polytetrafluoroethylene (PTFE) membrane. The material allows air to pass through relatively easily yet can withstand water exceeding 80 psi from entering the enclosure.

2.8 Life Cycle Battery Testing

This SAE Recommended Practice, J2288-2008, defines a standardized test method to determine the expected service life, in cycles, of electric vehicle battery modules [13]. It is based on a set of nominal or baseline operating conditions in order to characterize the expected degradation in electrical performance as a function of life and identify-relevant failure mechanisms where possible. Accelerated aging is not included in the scope of this procedure, although the time compression resulting from continuous testing may unintentionally accelerate battery degradation unless test conditions are carefully controlled. Because the intent is to use standard testing conditions whenever possible, results from the evaluation of different technologies should be comparable. End-of-life is determined based on module capacity and power ratings. This may result in a measured cycle life different than that which would be determined based on actual capacity; however, this approach permits a battery manufacturer to make necessary tradeoffs between power and energy in establishing ratings for a battery module. This approach is considered appropriate for a mature design or production battery. It should be noted that the procedure is functionally identical to the USABC Baseline Life Cycle Test Procedure.

This SAE Recommended Practice J2464_200911 "Electric and Hybrid Electric Vehicle Rechargeable Energy Storage System (RESS) Safety and Abuse" describes a tests that may be used as needed for abuse testing of electric or hybrid electric vehicle rechargeable energy storage systems (RESS) to determine the response of such electrical energy storage and control systems to conditions or events that are beyond their normal operating range [14]. These abuse test procedures are intended to cover a broad range of vehicle applications as well as a broad range of electrical energy storage devices, including individual RESS cells (batteries or capacitors), modules, and packs. This applies to vehicles with RESS voltages above 60 V.

Among the battery abuse tests are:

- Nail penetration
- Overcharge/overdischarge
- Short circuit
- Thermal stability
- Overtemperature
- Cycling without active cooling
- Crush/crash
- Drop
- Fire

2.9 Solid-State Batteries

Known commonly as solid-state batteries, solid-electrolyte batteries are believed to be the next major step in battery technology. They would offer superior safety to liquid-electrolyte batteries because of their lesser chance of a fire when their cases are punctured and offer better energy density, which could translate into smaller, lighter batteries [15].

CEO for Panasonic North America, Tom Gebhardt, said that solid-state batteries for electric vehicles are at least a decade out for mass-market electric vehicles [6]. A solid-state battery with two to four times the energy density would provide significant economic and cost advantages, as well as other secondary effects. However, most current solid-state battery designs are limited by the difficulty in processing the ceramic electrolyte components, which requires high temperatures that lead to material incompatibilities and high-energy costs.

Gebhardt said that, like all technologies, the industry loves to exaggerate the next major development in the battery field, no matter how far out. He expects current lithium-ion battery technology to undergo gradual refinement, improving energy density, charge rate, safety, and cost-effectiveness.

"We're still pretty bullish on lithium ion but clearly understand that solid state is something that we all want to get to at some point in the future," Gebhardt said.

Experts disagree as to whether lithium-ion batteries can become much cheaper. *Bloomberg* analysts predict a reduction in lithium-ion battery prices by as much as 52% by 2030 [17], while one BMW board member insists battery prices can't fall far below where they are today, at $116–$174 per kilowatt-hour of storage. Some companies are pouring money into solid-state battery development with the hope of readying the technology by 2025, while others believe lithium-sulfur batteries could serve as an intermediate solution between lithium-ion and solid-state batteries.

Researchers at Australia's Deakin University say they've managed to create a solid-state Li-ion battery. So far they've proven the process in coin cell batteries, similar to a watch battery size. The next step is to scale-up the batteries for higher power applications.

Dr. Fangfang Chen and Dr. Xiaoen Wang claim to have made a breakthrough with "the first clear and useful example of liquid-free and efficient transportation of lithium-ion in the scientific community." The new technology uses a solid polymer material weakly bonded to the lithium-ion to replace the volatile liquid solvents typically used as electrolytes in current battery cells. The liquid electrolyte is the part of the Li-ion battery that can become flammable.

Besides making batteries safer, the team believes this solid polymer electrolyte will allow batteries to work with a lithium metal anode. Dr. Wang says this could double the energy density of lithium batteries, which, in commercial settings, are currently peaking at around 250 Wh/kg [18].

2.10 Battery Thermal Management

The EV's battery also has thermal management issues. Heat is generated in the battery pack when charging and discharging current and the internal resistance of the battery cells and interconnections during vehicle acceleration and deceleration. Like the motors and power electronics components, the EV's battery is sensitive to operating temperature [19].

From a thermal point of view, there are three main aspects to consider when using lithium-ion batteries in an EV:

1. At temperatures below 0°C, batteries lose charge due to slower chemical reactions taking place in the battery cells. The result is a significant loss in power, acceleration, and driving range, and higher potential for battery damage during charging.
2. At temperatures above 30°C the battery performance degrades, posing a real issue if a vehicle's air conditioner is needed for passengers. The result is an impact on power density and reduced acceleration response.
3. Temperatures above 40°C (104°F) can lead to serious and irreversible damage in the battery. At even higher temperatures, for example, 70°C–100°C, thermal runaway can occur. This is triggered when the runaway temperature is reached. The result is a self-heating chain reaction in a battery cell that causes its destruction while propagating to adjacent cells.

The ideal temperature range for an EV's lithium-ion battery is similar to that preferred by human beings. To keep it in this range, the battery temperature must be monitored and adjusted. A battery thermal management system (BTMS) is necessary to prevent temperature extremes, ensure proper battery performance, and achieve the expected life cycle. An effective BTMS keeps cell temperatures within their allowed operating range.

2.11 Battery Chargers

In home charging, the EV connects to the power grid through standard socket outlets, which, depending on the country, is usually rated at around 10 A at 120 VAC. The electrical installation must comply with the region's safety regulations and needs a grounding system and a circuit breaker. There are two types of chargers: external to the EV (off-board) and onboard the EV.

Onboard chargers are buried within the car. These chargers accept AC power from the home and charge the car's battery. Onboard and off-board battery chargers can use a wall-mounted box that supplies 240 VAC to the car's charger. That box, cord, and plug is called the electric vehicle service equipment (EVSE). A 30-A EVSE will need a circuit breaker rated for at least 40 A.

An EVSE should handle at least 30 A. A rule of thumb is that a 30 A service will provide the ability to add about 30 miles of range in an hour. In contrast, 15 A will add only about 15 miles in an hour of charging. A cable from the wall-mounted EVSE to the automobile usually runs about 15–25 ft.

Some EV owners want ultrafast charging, which can be done, but it should be used sparingly as fast charging stresses the battery. Ultrafast charging is ideal for EV drivers on the run and this is fine for occasional use. Some EVs keep a record of stressful battery events and this data could be used to nullify a warranty claim. If at all possible, do not exceed a charge rate of 1C. Avoid full charges that take less than 90 min.

Onboard and off-board EV chargers must use switch-mode technology to provide high power and efficient operation. When the charger is connected to an AC power line, the switch-mode battery charger causes current and voltage distortions on the power line. This distortion can affect the operation of other equipment connected to the same power line.

Power line distortion is related to power factor, which is the ratio of real (useful) power in watts to apparent power in volt-amperes (VA Reactive). Current and voltage distortion (called harmonic distortion) can occur with a reactive load, which has two components: real and reactive. The vector sum of these two power components is the apparent power applied to the load. The phase angle between the real power and reactive power is called the power factor. With a resistive load, the reactive power is zero and the apparent power equals the real power and the power factor is 1. If the load is reactive, the power factor is less than 1. Reactive current doesn't do any real work, but it does incur real power losses in the resistance of the wiring, and these losses must be offset by the utility increasing its generating capacity.

To mitigate the effect of this current/voltage distortion, chargers incorporate a power factor correction (PFC) circuit, which performs as implied, ahead of the battery charger. Most PFC circuits use controller ICs to simplify the design, reduce circuit complexity, and minimize cost and weight. Besides correcting the power line voltage and current for the charger circuit, the PFC employs a transformer for safety purposes to isolate the battery and associated circuits from the AC power line.

2.12 Onboard Charger Details

An onboard charger (OBC) enables charging of the battery from the AC mains either at home or from outlets found in private or public charging stations. These chargers range from a 3.6 kW single-phase to a 22 kW three-phase, high-power converter. These OBCs must have high possible efficiency and reliability to ensure rapid charging times as well as meet the limited space and weight requirements.

These OBCs can employ power semiconductors that are AEC-Q101 qualified IGBTs, silicon, and SiC MOSFETs and diodes. AEC-Q100 qualified galvanically isolated IGBT and MOSFET gate drivers and automotive microcontrollers are also used. Newer chargers may use silicon carbide (SiC) or gallium nitride (GaN) power semiconductors.

Table 2.2 lists the specifications for the CurrentWays Technologies CWBC-Series of EV onboard, liquid-cooled chargers [20]. This onboard charger operates from 240

TABLE 2.2 Specifications for the current ways CWBC-series onboard charger.

Input		Output	
Voltage	90–264 VAC	Rated power	6.6 kW at 240 VAC
Current	19 A/32 A	Output voltage	250–450 VDC
Frequency	47–63 Hz	Ripple current	<2% below 20 MHz
Inrush current	<40 A at 240 VAC	Set point accuracy	±2%
Protection	Internally fused. 40 A both lines	Over current protect	110% self-limiting
Leakage	<75 mA at 240 VAC, 60 Hz	Overvoltage protect	120% threshold
Harmonics	EN61000-3-2	Overpower protect	115% threshold
Environmental			
Protection	IP67 rated against water and/or dust (submersible for 30 min at 1 m)		
Thermal shake and vibration	Designed to meet GMW-3172		
Operating temperature	-20°C to +65°C input liquid temperature without derating (with 60/40 mix glycol/water coolant)		
Over temperature protection	85°C internal temperature		
Cooling	Liquid cooling 9 L/min		
EMI compliance	Designed to meet CSPR 25		
Safety compliance	Designed to meet UL2202, J-3072, Model-based design		

VAC (6.6 kW) or 120 VAC (1.8 kW). It includes two CAN ports and is compliant with ISO 14229 Unified Diagnostic Services. Size is 400 × 255 × 107 mm and weight is 11 kg. Two brass fittings at the bottom of the unit are the inlet and outlet of the water-cooling valves.

2.13 **Smart Chargers**

Charging EV batteries may eventually pose a challenge to the nation's electricity grid as more EVs hit the road. It's possible that uncontrolled charging of EVs at maximum power might require substantial, costly investments in the nation's electricity system. And, enlarging a utility to meet increased peak demands could lead to excess generating capacity that will sit idle much of the time.

One answer to variable grid loading is smart chargers that control their demand depending on grid loads and the vehicle owner's needs. This gets complicated because grid loads are variable. Therefore, a smart charger must shift battery charging loads to run when there is less power demand on the grid. This requires the EV to send the vehicle's battery SoC to the utility. Without it, vehicle owners face the risk of not charging the battery properly.

Smart charging must be tailored to the types of charging stations that vary in their location and usage:

- Typically, for a residential charging station the EV owner plugs in when returning home and the car can charge overnight. A home charging station usually has no user authentication, no metering, and may require wiring a dedicated circuit.
- Charging in public charging stations may be slow, so it encourages EV owners to charge their cars while they are taking advantage of nearby facilities.

- To allow longer distance trips, there may be fast charging at public charging stations with >40 kW capability.

Public charging stations may be networked, that is, several could be linked together at multiple locations, and the utility receives information from all of them. These networked EV charging systems require "intelligent charging solutions," where chargers are linked seamlessly with the grid and charging system operators.

To aid the control of power required from the grid:

- "Smart chargers" should adjust their charging rate to accommodate variations in grid power demand.
- The electric grid should see a gradual "ramp" from the mid-day load to the evening peak

2.14 Thermal Management of Onboard Chargers

Plug-in electric vehicles (PEVs) present some unique thermal challenges [21]. While battery thermal management deals with bulk heat removal, the power electronics requires heat removal from tightly packed, concentrated heat loads [22].

One of the key challenges in PEVs is the time required for charging the batteries and the availability of power outlets. Charging of PEVs is classified into *Levels 1*, *2*, and *3* by the SAE (Table 2.3).

The advantage of having an onboard charger (versus off-board) is that the vehicle can be charged from AC power outlets. However, it also requires the vehicle to carry the extra weight of the power electronics and heat sinks.

High-power levels dissipated by chargers requires attention to their thermal management. Therefore, it's a good idea to estimate the power dissipation from the electronic components in the charger. The overall power conversion efficiency (AC to DC) is usually in the range of 93%–94%, which means that a 3.3 kW charger (Level 1 charging at 220 V, 15 A) will dissipate around 250 W. Integrated, multifunctional chargers will have additional power loads due to DC–DC conversion. However, since the AC–DC load (charging mode) and DC–DC load (drive mode) do not occur at the same time, the loads do not add up. This makes the thermal management of a multifunctional charger easier in being able to share a common heat sink, which reduces overall size, weight, and cost.

The OBC electronics need to be packaged inside an enclosure that is sealed to prevent environmental contamination. This requires the heat loads be thermally connected to

TABLE 2.3 SAE charging levels.

Level	Description
Level 1	Slow charging at 120/240 V AC and 15 A using the standard available household power outlets up to 3.3 kW. The AC to DC power conversion is done onboard.
Level 2	Medium rate charging using 240 V AC and 60 A up to 14.4 kW from power outlets specifically made for PEV charging. The AC to DC power conversion is done on board.
Level 3	Fast charging used specially for PEV charging rated over 14.4 kW. In this case, the AC to DC power conversion is usually done off-board.

the enclosure walls in order to efficiently dissipate the heat. The enclosure wall, therefore, needs to function as a heat sink to dissipate the heat to the outside air (or liquid). To ensure that the heat loads are thermally connected to the enclosure wall, a suitable thermal-interface material, which provides not only good thermal conduction but also the required electrical insulation between the device and the enclosure, must be selected.

For a relatively low power case, the heat load can be easily removed by forced air convection from the outside of the enclosure walls using a fan. Considering that the charging load occurs when the vehicle is stationary, there is no additional airflow benefit due to vehicle movement. The systems integrator has great flexibility on how to position the charger inside the vehicle. The weight of the charger can be significantly reduced by adding heat pipes to the heat sink base to spread the heat from the concentrated heat loads.

2.15 Wireless Battery Charging

Wireless charging may one day replace plugs and wires used to charge an EV battery. Wireless charging sends signals in a near field condition in which the primary coil produces a magnetic field that is induced into a secondary coil that is in close proximity. Most of today's wireless chargers use inductive charging with transmit and receive coils in close proximity. Larger batteries for an EV use resonance charging by making a coil "ring." The oscillating magnetic field works within a 1-m (3-ft) radius. To stay in the power field, the distance between transmit and receive coil must be within a quarter wavelength (915 MHz has a wavelength of 0.328 m or 1 ft).

Resonance charging can also be used at all power levels. While a 3 kW system for EV charging achieves a reported efficiency of 93%–95% with a 20 cm (8 in.) air gap, a 100 W system is better than 90% efficient; however, the low power, 5 W system remains in the 75%–80% efficiency range. Resonance charging is still in the experimental stages and is not widely used.

Modern wireless charging follows a complex handshake to identify the device to be charged. When placing a device on a charge mat, the change in capacitance or resonance senses its presence. The mat then transmits a burst signal; the qualified device awakens and responds by providing identification and signal strength status.

A charge mat only transmits power when a valid object is recognized, which occurs when the receiver fulfills the protocol as defined by one of the interoperability standards. During charging, the receiver sends control error signals to adjust the power level. Upon full charge or when removing the load, the mat switches to standby.

Transmit and receive coils are shielded to obtain good coupling and to reduce stray radiation. Some charge mats use a free-moving transmit coil that seeks the object placed for best coupling. Other systems feature multiple transmit coils and engage those in close proximity with the object.

BMW wireless charging enables electric energy from the mains supply to be transmitted to a vehicle's HV battery, without any cables, when the vehicle is positioned over a base pad [23]. This can be installed in the garage, for example, and the charging process started as soon as the vehicle is parked in position (without any further input from the driver). The launch of this technology sees the BMW Group move another step closer to an infrastructure that will make charging the battery of an electrified vehicle even simpler than refueling a car with a conventional engine.

Available as an option, BMW's wireless charging consists of an inductive charging station (GroundPad) that can be installed either in a garage or outdoors and a secondary vehicle component (CarPad) fixed to the underside of the vehicle. The contactless transfer of energy between the GroundPad and CarPad is conducted over a distance around 8 cm. The GroundPad generates a magnetic field that is induced in the CarPad and then charges the HV battery.

The system has a charging power of 3.2 kW, enabling the HV batteries onboard the BMW 530e to be fully charged in about 3½ h. And with an efficiency rate of around 85%, charging with the BMW wireless charging system is efficient.

BMW wireless charging employs the same inductive charging technology already widely used for supplying power to devices, such as mobile phones and electric toothbrushes, to now recharging the HV batteries in electrified vehicles. The principal benefit here is the ease of use, because drivers no longer need to hook up their plug-in hybrid car using a cable in order to replenish its energy reserves. Instead, as soon as the vehicle has been parked in the correct position above the inductive charging station, followed by a simple push of the start/stop button, the charging process is initiated. Once the battery is fully charged, the system switches off automatically.

BMW wireless charging also helps the driver to maneuver into the correct parking position. A Wi-Fi connection establishes communication between the charging station and vehicle. The EV's control display presents colored lines that help guide the driver while parking. A graphic icon shows when the correct parking position for inductive charging has been reached. This can deviate from the optimum position by up to 7 cm longitudinally and up to 14 cm laterally.

References

1. Wikipedia, "Electric Vehicle Battery."
2. Cadex Electronics, "Electric Vehicles (EV) BU-1003," Cadex Electronics Battery University (Update), January 28, 2019.
3. Cadex Electronics, "Is Li-Ion the Solution for the EV?," Cadex Electronics, January 2011.
4. MIT, "A Guide to Understanding Battery Specifications," MIT Electric Vehicle Team, December 2008.
5. Hu, R., "Battery Management System for EV Applications," University of Windsor, September 16, 2011, 1-82.
6. Cadex Electronics, "BU-908, Battery Management System," Cadex Electronics, April 24, 2019.
7. Renesas, "Battery Management System Tutorial," Renesas, August 2018.
8. DOE, "Batteries for Hybrid or Plug-In Electric Vehicles," Alternative Fuels Data Center (DOE).
9. Cadex Electronics, "Cell Matching and Balancing," BU-803a, Cadex Electronics, Battery University, April 2, 2016.
10. Texas Instruments, "bq76PL455A-Q1 16-Cell EV/HEV Integrated Battery Monitor and Protector," Texas Instruments Data Sheet, November 11, 2016, 1-130.
11. Stephens, D., Shawcross, P., Stout, G., Sullivan, E. et al., "Lithium-Ion Battery Safety Issues for Electric and Plug-In Hybrid Vehicles," National Highway Traffic Safety Administration, October 12, 2017.
12. Clemens, K., "Two-Stage Venting Provides Battery Protection," Design News, June 19, 2018.

13. SAE, "Life Cycle Testing of Electric Vehicle Battery Modules," J2288_200806, June 30, 2008.
14. Doughty, D., "SAE, J2464 EV & HEV Rechargeable Energy Storage System (RESS) Safety and Abuse Testing Procedure," SAE Technical Paper 2010-01-1077, 2010, https://doi.org/10.4271/2010-01-1077.
15. Gilboy, J., "Solid State EV Batteries at Least a Decade Away," The Drive, November 14, 2018.
16. Davis, S., "Who Will Win the Race for an EV Solid State Battery?," powerelectronics.com, August 24, 2018.
17. Bullard, N., "Electric Car Price Tag Shrinks Along With Battery Cost," Bloomberg, April 12, 2019.
18. Wang, X., Chen, F. et al., "Poly(Ionic Liquid)s-in-Salt Electrolytes with Co-coordination-Assisted Lithium-Ion Transport for Safe Batteries," Joule, August 7, 2019.
19. Quesnel, N., "Industry Developments in Thermal Management of Electric Vehicle Batteries," Advanced Thermal Solutions, October 12, 2018.
20. CurrentWays, "Multi-Stage Electric Vehicle Battery Charger," CurrentWays.
21. Continental, "Battery Management System," Continental.
22. Goswami, A., "Thermal Management of On-Board Chargers," Electronics Cooling, August 24, 2017.
23. Lambert, F., "BMW Launches Wireless Electric Car Charging System Touted as Convenient but Inefficient," Electrek, May 28, 2018.

CHAPTER 3

EV Fuel Cells

A fuel cell employs an electrochemical process to generate DC power, which allows it to replace the traction battery in an EV. In operation, the fuel cell converts up to 60% of its chemical energy to power a vehicle. A fuel cell in an electric vehicle is potentially three times more efficient than a traditional combustion vehicle. In addition, it is virtually pollution-free [1, 2].

3.1 Fuel Cell Basics

Fuel cell EVs achieve the beneficial characteristics of both conventional and battery EVs. The combination of these desirable characteristics makes fuel cell electric vehicles an attractive advanced vehicle option for drivers that value long range, fast refueling, and zero emissions.

The power produced by a fuel cell depends on:

- Fuel cell type
- Size
- Operating temperature
- Pressure of supplied gases

A single fuel cell produces roughly 0.5–1.0 V, barely enough voltage for even the smallest applications [3]. To deliver the desired amount of energy, the fuel cells must be combined in series to yield higher voltage and in parallel to supply higher current. These series and parallel combinations produce a fuel cell stack. The cell surface area can also be increased to allow higher current from each cell. Within the stack, reactant gases must be distributed uniformly over each of the cells to maximize the power output. Depending on the application, a fuel cell stack may contain only a few or as many as hundreds of individual cells layered together. This "scalability" makes fuel cells ideal for a wide variety of applications, from vehicles (50–125 kW) to laptop computers (20–50 W), homes (1–5 kW), and central power generation (1–200 MW or more).

A typical fuel cell output voltage decreases as current increases, due to several factors:

- Activation loss
- Ohmic loss (voltage drop due to resistance of the cell components and interconnections)
- Mass transport loss (depletion of reactants at catalyst sites under high loads, causing rapid loss of voltage)

Fuel cell electric vehicles (FCEVs) are much cleaner than gasoline and diesel-powered automobiles [4]. When gasoline or diesel fuels are burned in a car or truck, smog and greenhouse gases (GHG) emissions are created. FCEVs are a "zero-emission vehicle," which means they generate no tailpipe emissions of harmful air pollutants or GHGs. FCEVs currently achieve slightly lower reductions in climate pollution than pure battery electric vehicles (BEVs). Therefore, the California Energy Commission (CEC) encourages new car buyers to carefully consider their day-to-day driving needs and the performance of BEVs and FCEVs before moving forward with the purchase of the vehicle [5]

3.2 Polymer Electrolyte Membrane Fuel Cell (PEMFC)

The most common type of fuel cell for EV applications is the PEMFC [6]. In the PEMFC, the polymer electrolyte membrane separates the anode and cathode sides of the fuel cell (Figure 3.1). The PEMFC allows only the positively charged ions to pass through it to

FIGURE 3.1 Polymer electrolyte membrane fuel cell.

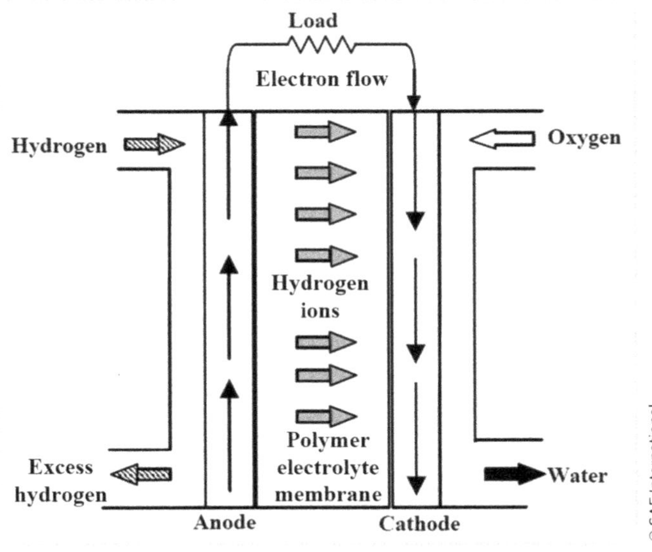

the cathode. The negatively charged electrons must travel along an external circuit to the cathode, creating an electrical circuit.

The hydrogen fuel is channeled through field flow plates to the anode on one side of the fuel cell, while an oxidant (oxygen or air) is channeled to the cathode on the other side of the cell. At the anode, a platinum catalyst causes the hydrogen to split into positive hydrogen ions (protons) and negatively charged electrons. At the cathode, the electrons and positively charged hydrogen ions combine with oxygen to form water, which flows out of the cell.

The materials used for different parts of the fuel cells differ by type. The bipolar plates may be made of different types of materials, such as metal, coated metal, graphite, flexible graphite, C–C composite, carbon–polymer composites, and so on. The PEM is sandwiched between two catalyst-coated carbon papers.

In the PEMFC, the membrane must be hydrated, requiring water to be evaporated at the same rate it is produced. If water is evaporated too quickly, the membrane dries, resistance across it increases, and it will eventually crack, creating a gas "short circuit" where hydrogen and oxygen combine directly, generating heat that will damage the fuel cell. If the water is evaporated too slowly, the electrodes will flood, preventing the reactants from reaching the catalyst and stopping the reaction. A steady ratio between the reactant and oxygen is necessary to keep the fuel cell operating efficiently.

3.3 Fuel Cell Thermal Management

PEMFCs have a technical challenge of thermal management [7]. Any mismanagement of water and/or heat could lead to water flooding or membrane dry-out, which affects fuel cell performance and durability. In addition, the PEMFC can only tolerate a small temperature variation.

Two factors are critical when designing a cooling system for PEMFCs:

1. The nominal operating temperature of a PEMFC is limited to roughly 80°C, so natural heat rejection is minimal.
2. Nearly the entire waste-heat load must be removed by an ancillary cooling system, because the exhaust streams contribute little to the heat removal.

A study to determine suitable cooling for a fuel cell system used simulation software to model the cooling system [8]. Then, both steady-state and transient simulations were performed. It was found that the cooling system can provide sufficient heat rejection for the current fuel cell system, even at demanding driving conditions up to ambient temperatures of at least 45°C. Further, for the more powerful fuel cell system the cooling system can only sustain sufficient heat rejection for less-demanding driving conditions, hence it was concluded that improvements were needed. The following improvements were suggested:

- Increase air mass flow rate through the radiator
- Increase pump performance
- Remove the heat exchanger in the cooling system

It was found that, if these improvements were combined, the cooling system could sustain sufficient heat rejection for the more powerful fuel cell system up to the ambient temperature of 32°C.

3.4 Fuel Cell Energy Efficiency

The energy efficiency of a system or device that converts energy is measured by the ratio of the amount of useful energy put out by the system (output energy) to the total amount of energy that is put in (input energy) or by useful output energy as a percentage of the total input energy. In the case of fuel cells, useful output energy is measured in electrical energy produced by the system. Input energy is the energy stored in the fuel. According to the U.S. Department of Energy, fuel cells are generally between 40% and 60% energy efficient, which is higher than the typical internal combustion engine of a car that is about 25% energy efficient. In combined heat and power (CHP) systems, the heat produced by the fuel cell is captured and utilized, increasing the efficiency of the system to up to 85%–90%.

The theoretical maximum efficiency of any type of a power generation system is never reached in practice, and it does not consider other steps in power generation, such as production, transportation and storage of fuel, and conversion of the electricity into mechanical power. However, this calculation allows the comparison of different types of power generation. The maximum theoretical energy efficiency of a fuel cell is 83%, operating at low power density and using pure hydrogen and oxygen as reactants (assuming no heat recapture). According to the World Energy Council, this compares with a maximum theoretical efficiency of 58% for internal combustion engines.

FCEVs are powered by hydrogen. They are more efficient than conventional internal combustion engine vehicles and produce no tailpipe emissions-they only emit water vapor and warm air. The U.S. Department of Energy is leading government and industry efforts to make hydrogen-powered vehicles an affordable, environmentally friendly, and safe transportation option. Hydrogen is considered an alternative fuel under the Energy Policy Act of 1992 and qualifies for alternative fuel vehicle tax credits. In FCEVs, hydrogen is stored in tanks and then converted to electricity (Figure 3.2).

FCEVs are fueled with pure hydrogen gas [9]. Similar to conventional internal combustion engine vehicles, they can fuel in less than five minutes and have a driving range over 300 miles. FCEVs are equipped with other advanced technologies to increase efficiency, such as regenerative braking systems, which capture the energy lost during braking and store it in a battery. Major automobile manufacturers are offering a limited

FIGURE 3.2 Fuel cell employed in an EV [1].

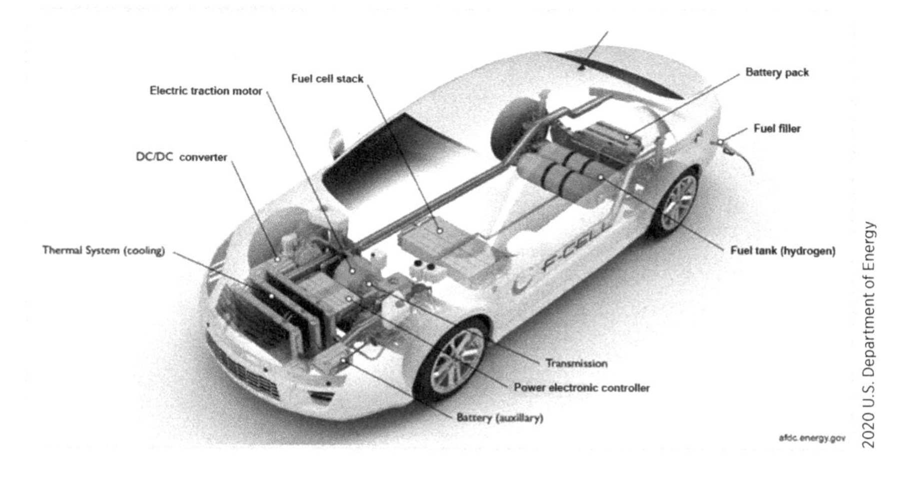

number of production FCEVs to the public in certain markets in sync with what the developing infrastructure can support.

As of 2015, two fuel cell vehicles were introduced for commercial lease and sale in limited quantities: the Toyota Mirai and the Hyundai ix35. Additional demonstration models include the Honda FCX Clarity and Mercedes-Benz F-Cell. General Motors and its partners estimated that per mile traveled, a fuel cell electric vehicle running on compressed gaseous hydrogen produced from natural gas could use about 40% less energy and emit 45% less GHG than an internal combustion vehicle. A lead engineer from the Department of Energy whose team is testing fuel cell cars in 2011 said the potential appeal is that, "these are full-function vehicles with no limitations on range or refueling rate, so they are a direct replacement for any vehicle."

3.5 Toyota Mirai

Toyota Mirai Specifications (2019) [10]:

- **Range per tank**: 502 km/312 mi
- **Curb weight**: 1,848 kg/4,075 lb
- **MPGe**: 66 City/66 Highway/66 Combined MPGe rating (EPA cycle)
- **Electric Motor Power**: 114 kW, 153 hp
- **Electric Motor Torque**: 247 lb-ft
- **Motor Type**: Permanent Magnet AC synchronous
- **Hydrogen tank capacity**: 122.4 L total (fore tank: 60 L, aft tank: 62.4 L)
- **Battery capacity**: 1.6 kW
- **Battery type**: Nickel-Metal Hydride (Ni-MH)

While the numbers will remain small compared with vehicles fitted with internal combustion engines, battery electric, or hybrid propulsion, automakers and suppliers are planning for substantial growth in the segment as the world's transportation systems slowly transform to a hydrogen-driven future.

A research report by Wards Intelligence, "FCVs On The Horizon," reveals a number of factors driving that growth: reductions in the cost and size of fuel cell stacks; improvements in fuel cell output and durability; public and private efforts to expand the fueling infrastructure and clean sources of hydrogen; improvements in electric motors, energy storage, and control systems; and the application of fuel cells to economically power everything from forklifts to vehicles.

A Wards Intelligence survey of OEMs and suppliers and projections from data partner LMC Automotive suggest fuel cell demand in light vehicles will pick up momentum by 2025, initially finding the greatest opportunities in full-size SUVs and trucks. Commercial vehicles, buses, heavy-duty trucks, and forklifts will provide significant growth opportunities as well.

Bob Gritzinger of Wards Intelligence said that fuel cell vehicles represent a tiny niche in the global automotive market, but their mere existence today as functional, everyday cars marketed by major automakers suggests the technology is on a path to wider proliferation in the coming decade [11].

"It is the first part of the curve, the first 100-1,000 units, that is most challenging," says Keith Wipke, laboratory program manager at the U.S. Department of Energy's

National Renewable Energy Laboratory. "If an automaker like Toyota is producing 30,000 (FCEVs) in 2020, then 100,000 in 2025 is reasonable" [9].

The main impediment to fuel cell use is the fueling infrastructure. There aren't enough stations, and hydrogen is expensive to produce. And, depending on the source of hydrogen—if it is produced from coal-fired electric power, for instance—it will not be clean. Technologically, while on-road operation has been validated and the stack has been downsized sufficiently to be placed in a compact car's engine compartment, system cost is still prohibitive, although a pathway can be seen to fill certain niches based on duty cycle and application.

Among the report's key findings:

- The current market leader in FCEVs, Toyota, will cede its dominance in the coming decade as Honda and Hyundai grow their FCEV portfolios.
- Europe will supplant Asia-Pacific in FCEV volume by 2030, joining North America in leading FCEV deliveries.
- In the big picture, the numbers still will remain tiny with about 301,000 annual FCEV sales in 2030, well under 1.0% of global volume. However, that number represents a major increase over the infinitesimal 575 FCVs sold in 2015 and signals that FCVs are on a path to wider proliferation in the 2030s and beyond.

3.6 Flexible Nanomaterials Fuel Cell

A new method of increasing the reactivity of ultrathin nanosheets, just a few atoms thick, can someday make fuel cells for hydrogen cars cheaper, as described in a Johns Hopkins study [12]. A report of the findings offers promise toward faster, cheaper production of electrical power using fuel cells, but also of bulk chemicals and materials such as hydrogen.

"Every material experiences surface strain due to the breakdown of the material's crystal symmetry at the atomic level. We discovered a way to make these crystals ultrathin, thereby decreasing the distance between atoms and increasing the material's reactivity," says Chao Wang, an assistant professor of chemical and biomolecular engineering at the Johns Hopkins University and one of the study's corresponding authors.

In this study, Wang and colleagues manipulated the strain effect, or distance between atoms, causing the material to change dramatically. By making those lattices incredibly thin, roughly a million times thinner than a strand of human hair, the material becomes much easier to manipulate, just like how one piece of paper is easier to bend than a thicker stack of paper.

"We're essentially using force to tune the properties of thin metal sheets that make up electrocatalysts, which are part of the electrodes of fuel cells," says Jeffrey Greeley, professor of chemical engineering at Purdue and another one of the paper's corresponding authors. "The ultimate goal is to test this method on a variety of metals."

"By tuning the material's thinness, we were able to create more strain, which changes the material's properties, including how molecules are held together. This means you have more freedom to accelerate the reaction you want on the material's surface," explains Wang.

One example of how optimizing reactions can be useful in application is increasing the activity of catalysts used for fuel cell cars. While fuel cells represent a promising technology toward emission-free electric vehicles, one challenge lies in the expense associated with the precious metal catalysts, such as platinum and palladium, limiting its viability to the majority of consumers. A more active catalyst for the fuel cells can reduce cost and clear the way for widespread adoption of green, renewable energy.

Wang and colleagues estimate that their new method can increase catalyst activity by 10–20 times, using 90% less of precious metals than what is currently required to power a fuel cell. "We hope that our findings can someday aid in the production of cheaper, more efficient fuel cells to make environmentally-friendly cars more accessible for everybody," says Wang.

Other study authors include Lei Wang, David Raciti, Michael Giroux of the Johns Hopkins University, Zhenhua Zeng and Tristan Maxson of Purdue University, and Wenpei Gao and Xiaoqing Pan of the University of California, Irvine.

3.7 Fuel Cell Hydrogen

Hydrogen's specific energy, the amount of energy it contains in a given weight, is far higher than that of lithium-ion batteries. One kilogram of hydrogen can store 236 times more energy than a 1-kg lithium-ion battery [2].

In the FCEV, hydrogen is stored in high-pressure, lightweight tanks underneath the vehicle. Energy stored in batteries requires many more battery cells for each extra mile added to the range of the vehicle. That means more weight that the vehicle has to overcome for all those miles. In terms of the weight of the energy storage in the vehicle, hydrogen clearly has the advantage, even when the fuel cell and its infrastructure are considered.

A paper by Staffel et al. presents a comprehensive review of the potential role that hydrogen could play in the provision of electricity, heat, industry, transport, and energy storage in a low-carbon energy system, and an assessment of the status of hydrogen in being able to fulfil that potential [13]. The picture that emerges is one of qualified promise: hydrogen is well-established in certain niches like forklift trucks, while mainstream applications are now forthcoming. Hydrogen vehicles are available commercially in several countries, and 225,000 fuel cell home-heating systems have been sold. This represents a steep change from the situation of only 5 years ago. This review shows that challenges around cost and performance remain, and considerable improvements are still required for hydrogen to become truly competitive. But such competitiveness in the medium-term future no longer seems an unrealistic prospect and fully justifies the growing interest and policy support for these technologies around the world.

Scientists are now testing a fuel cell-powered Toyota Mirai that uses ultrahigh purity hydrogen produced using a membrane technology developed by CSIRO, the Commonwealth Scientific and Industrial Research Organization, an independent Australian federal government agency responsible for scientific research [14].

Hydrogen is difficult to transport and store. Gaseous hydrogen can be transported by pipeline, but it tends to damage steel and needs considerable pipe wall thickness to ensure it does not escape. Hydrogen is very flammable and difficult to ship because of its low density. These logistical issues have always been a stumbling block. To produce hydrogen, CSIRO uses ammonia that can be stored at room temperature and easily converted into hydrogen by passing it over a catalyst to release hydrogen and nitrogen gas. CSIRO has successfully road tested its ammonia-to-hydrogen technology for a fuel cell-powered Toyota Mirai.

The key to the CSIRO project rests in a different approach based on a proprietary membrane separator technology. Its vanadium alloy membrane is tipped to transform the hydrogen separation process and enable the use of ammonia as a means of carrying the ultralight hydrogen. The thin metal membrane allows hydrogen to pass while blocking all other gases and, using decomposed ammonia feedstock, it enables hydrogen conversion in a single step. It permits a small plant—with no moving parts—to work in continuous operations.

3.8 Fuel Cell Electric Vehicle Evaluation

NREL's technology validation team analyzed hydrogen FCEVs operating in a real-world setting to identify the current status of the technology, compare it to Department of Energy (DOE) performance and durability targets, and evaluate progress between multiple generations of technology, some of which will include commercial FCEVs for the first time [9].

Current fuel cell, electric vehicle evaluations build on the 7-year FCEV learning demonstration and focus on fuel cell stack durability and efficiency, vehicle range and fuel economy, driving behavior, maintenance, onboard storage, refueling, and safety.

NREL is analyzing data from a DOE-sponsored demonstration project that includes six original equipment manufacturers—GM, Mercedes-Benz, Hyundai, Nissan, Toyota, and Honda—and up to 90 vehicles.

Participating partners share raw data with NREL via the National Fuel Cell Technology Evaluation Center. NREL engineers perform uniform analyses on the detailed data and then report on their findings. While the raw data is secured by NREL to protect proprietary information, individualized data analysis results are provided as detailed data products to the partners who supplied the data. The analysis results are also aggregated into publicly available composite data products that show the status and progress of the technology but don't identify individual companies.

References

1. AFDC, "Fuel Cell Electric Vehicles," Alternate Fuels Data Center (AFDC).
2. Drive Clean, "Hydrogen Fuel Cell," Drive Clean.
3. DoE, "Fuel Cells," DoE, November 2015.
4. Community Environmental Council, "Why Fuel Cell Electric Vehicles Matter," Community Environmental Council.
5. Woodford, C., "Fuel Cells," Explain that Stuff, updated: March 15, 2020.
6. Wikipedia, "PEFMC Fuel Cell."
7. Watzenig, D. and Brandstatter, B., Comprehensive Energy Management-Safe Adaptation, Predictive Control and Thermal Management, Springer Briefs in Applied Sciences and Technology (Cham: Springer International Publishing, 2018).
8. Swedenborg, S., "Modeling and Simulation of Cooling System for Fuel Cell Vehicle," Uppsala University, July 5, 2017.
9. Kurtz, J., Sprik, S., Saur, J., and Onorato, S., "Fuel Cell Electric Vehicle Durability and Fuel Cell Performance," NREL Technical Report, March 2019.
10. Toyota, "Mirai Fuel Cell Tech eBrochure," Toyota eBrochure, 2019, 1-22.
11. Gritzinger, R., "Fuel-Cell Vehicles to Pick Up Pace Mid-Decade, Study Concludes," Wards Intelligence, February 5, 2019.
12. Wang, L., Zhenhua, Z., Wenpei, G., Wang, C. et al., "Tunable Intrinsic Strain in Two-Dimensional Transition Metal Electrocatalysts," Science 363, no. 6429 (2019): 870-874.
13. Staffel, I., Scamman, D., Velazquez, A., Abad, B. et al., "The Role of Hydrogen and Fuel Cells in the Global Energy System," Royal Society of Chemistry 12 (2019): 463-491.
14. Ginn, E., "CSIRO Tech Accelerates Hydrogen Vehicle Future," CSIRO, August 6, 2018.

ECU Power Management

Power management for ECUs involves the use of power supplies to provide a regulated voltage for the ECU's electronic circuits. These power supplies accept a DC input from the main DC-DC converter and provide a DC output that is the same, higher, or lower than the input. The power supplies provide a regulated, constant output voltage [1]. These power supplies are referred to by several different names, which are used interchangeably:

- DC–DC converters
- Power converters
- Voltage regulators

4.1 Linear Power Supplies

There are two types of regulated power supplies employed in ECUs: linear and switch modes. Linear regulators operate in the linear region with the topology shown in Figure 4.1. Its main components are a series pass transistors (bipolar transistor or MOSFET), error amplifiers, and precise voltage references. The series pass transistor is always turned on, so that it always consumes power.

Most linear regulators used in the ECUs are low dropout (LDO) integrated circuits (ICs). Key operational factors for an LDO are its dropout voltage, power supply rejection ratio (PSRR), and output noise. Low dropout refers to the difference between the input and output voltages that enable the LDO to regulate the output load voltage. That is, an LDO will regulate the output voltage until its input and output approach each other at the dropout voltage. Ideally, the dropout voltage should be as low as possible to minimize

FIGURE 4.1 Basic LDO.

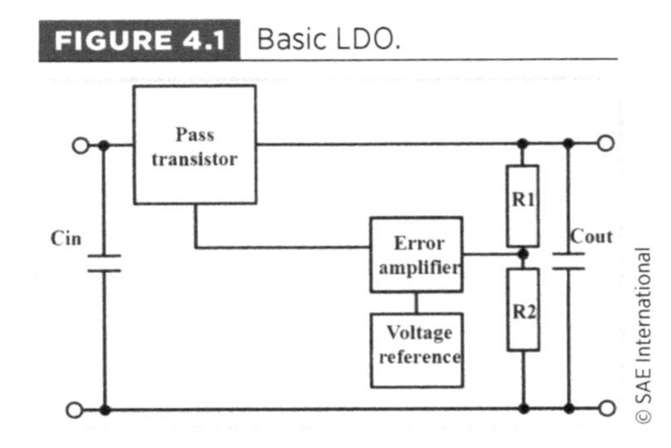

power dissipation and maximize efficiency. The load current and pass transistor temperature affect the dropout voltage. Efficiency of an LDO power supply is typically 40%–60% [2, 3].

4.2 Switch-Mode Power Supplies

A typical switch-mode power supply employs a pulse width modulator (PWM), power transistor switch, and error amplifier [4–6]. The PWM controls the frequency and duty cycle (on-and-off time) of the power transistor switch (Figure 4.2). Therefore, the power transistor switch produces a pulsed output signal. The power transistor switch consumes its maximum power when it is turned on and consumes very little power during its off time. By rectifying the pulsed output of the power transistor switch and filtering it with an inductor and capacitor low-pass filter, the supply produces a DC output. Rectifying and filtering the pulsed waveform also produces a small amount of output ripple on the DC output. The ripple is usually in the millivolt or microvolt range (Figure 4.3) [7].

Power supply efficiency is enhanced because the duty cycle of the power transistor switch is typically turned on only around 40%–60% of the time, depending on its output voltage. This switching configuration can provide a power supply with up to 95% efficiency (output/input).

To regulate the power supply's output voltage, a portion of the supply's DC output voltage is fed back to the error amplifier where it is compared with a fixed voltage reference (V_{REF}). If the output voltage feedback tends to increase relative to V_{REF}, the error amplifier causes the PWM circuit to reduce its duty cycle, causing the output voltage to reduce and maintain the proper regulated output voltage. Conversely, if the

FIGURE 4.2 PWM controller produces square waves of different widths dependent on the out voltage feedback.

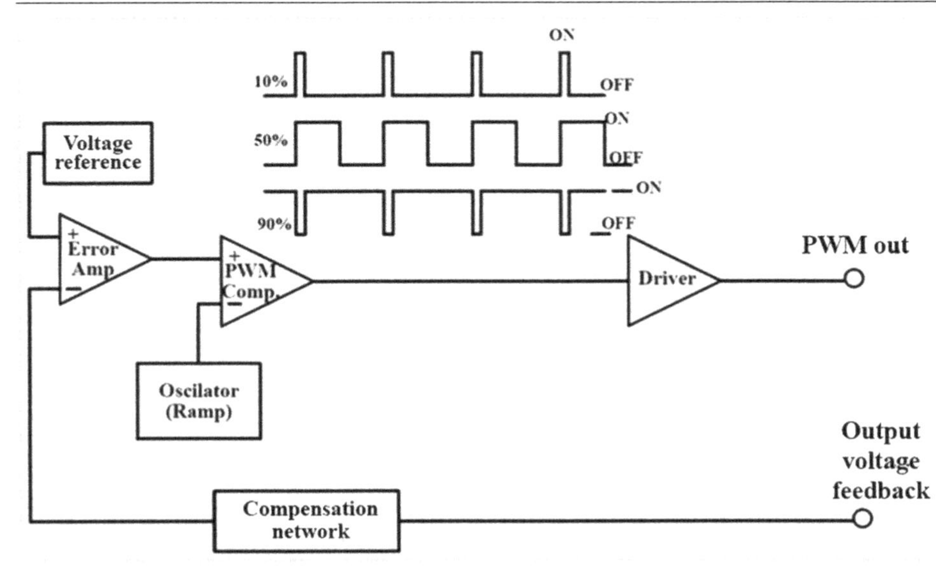

FIGURE 4.3 Switch-mode converter uses pulse width modulator to control regulation.

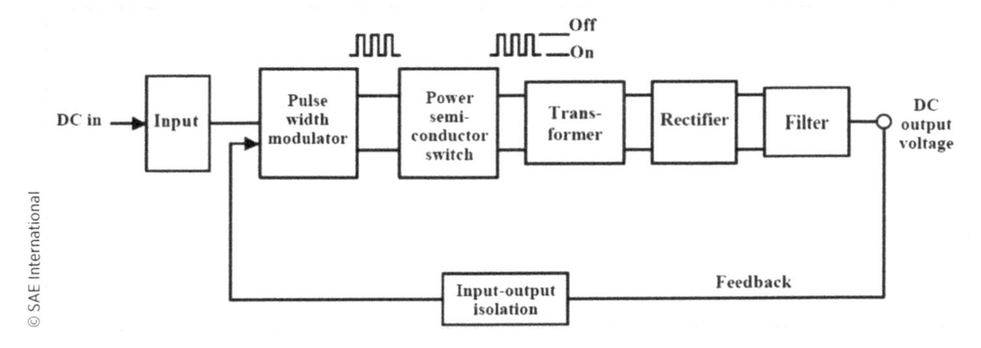

output voltage feedback tends to decrease, the error amplifier causes the PWM duty cycle to increase keeping the regulated output at its proper voltage.

Typically, the power transistor switch turns on and off at a frequency that may range from 100 kHz to over 1 MHz, which depends on the regulator IC that is used. Switching frequency determines the physical size and value of filter inductors and associated capacitors. That is, the higher the switching frequency the smaller their size and vice versa. To optimize efficiency, the inductor's magnetic core material should operate efficiently at the switching frequency. Additionally, the inductor core material should not saturate at the operating current.

There are two types of switch-mode converters: isolated and non-isolated, which depends on whether there is a direct DC path from the input to the output. An isolated converter employs a transformer to provide isolation between the input and output voltages (Figure 4.4). The non-isolated converter usually employs an inductor and there is no voltage isolation between the input and output (Figure 4.5). Some applications require isolation between the input and output voltages, but others do not.

Some converters employ a transformer that provides the ability to easily produce more than one output voltage simultaneously. In contrast, the inductor-based converter provides only one output voltage, depending on the circuit characteristics.

Table 4.1 compares the characteristics of linear and switch-mode supplies for use in ECUs.

FIGURE 4.4 Isolated switch-mode converter employs a transformer for isolation.

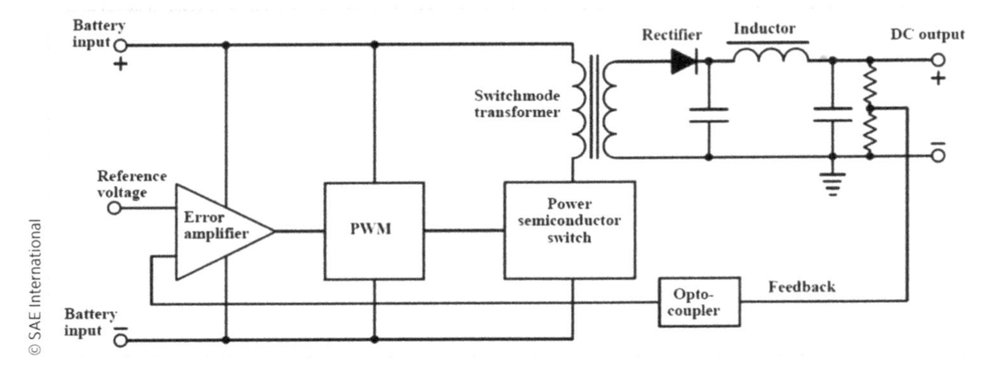

CHAPTER 4

FIGURE 4.5 Non-isolated switch-mode converter.

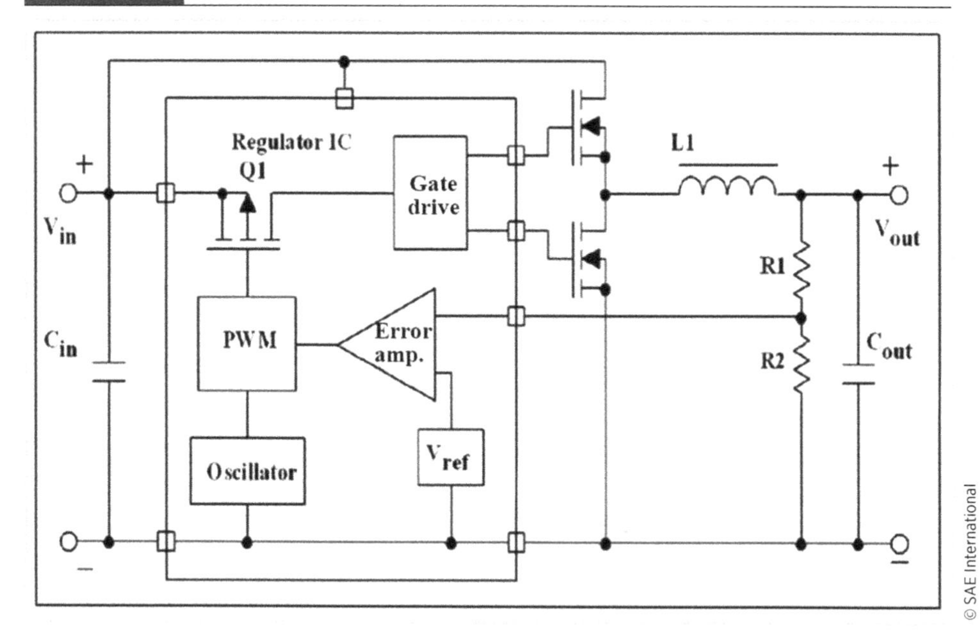

TABLE 4.1 Linear vs. switch-mode power supply comparison.

Parameter	Linear power supply	Switch-mode power supply
Efficiency	40%-60%	80%-95%
Power	Always on	Switched on and off
Design complexity	Simpler	More complex
Electromagnetic interference (EMI)	"Quiet" (None)	More (depends on switching frequency and layout)

4.3 Synchronous Rectification

Efficiency is an important criterion in designing DC–DC converters, which means power losses must be minimized [8]. These losses are caused by the power switch, magnetic elements, and the output rectifier. Reduction in power switch and magnetics losses require components that can operate efficiently at their switching frequency.

Output rectifiers can use Schottky diodes, but synchronous rectification (Figure 4.6) consisting of power MOSFETS can provide higher efficiency. MOSFETs exhibit lower forward conduction losses than Schottky diodes. Unlike conventional diodes that are self-commutating, MOSFETs turn on and off by means of a gate control signal synchronized with converter operation. The major disadvantage of synchronous rectification is the additional complexity and cost associated with the MOSFET devices and associated control electronics. At low output voltages, however, the resulting increase in efficiency more than offsets the cost disadvantage in many applications.

There are several possible topologies that can be used for switch-mode power supply circuits, as listed in Table 4.2.

Table 4.3 lists other important power supply characteristics that can affect ECU performance.

FIGURE 4.6 Synchronous rectifier is more efficient than a diode rectifier.

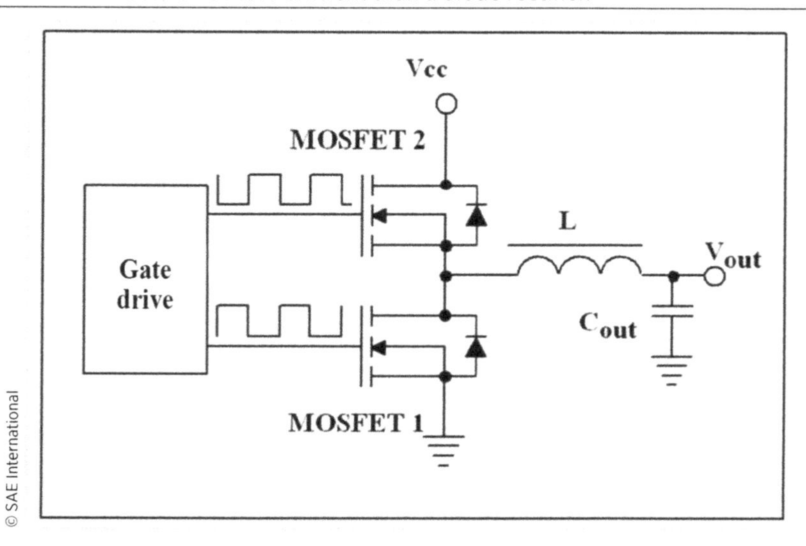

TABLE 4.2 Switch-mode power supply topologies.

Regulator topology	Description
Buck [9-13]	Buck converters step down the input voltage to a lower value output. They feature a controller with one or more integrated FETs and an inductor, and offer a balance between flexibility and ease of use. The inductor allows the flexibility to optimize the power supply for efficiency, size, or cost.
Boost [14, 15]	Steps up, or boosts, the input voltage to a higher value output. It has the advantages of simplicity, low cost, and the ability to step-up the output without a transformer. Disadvantages are a limited power range and a relatively high output ripple due to the off-time energy coming from the output capacitor.
Buck-Boost[16, 17]	Buck-boost converters combine elements of a buck and boost converter. The buck-boost converter is a switch-mode power supply that provides a regulated DC output from a source voltage either above or below the desired output voltage. This can be useful in battery-powered applications where the battery voltage starts out above the desired output but falls below as the battery drains.
SEPIC [18, 19]	The single-ended primary inductance converter (SEPIC) is a DC/DC converter topology that provides a positive regulated output voltage from an input voltage that varies from above to below the output voltage.
Hysteretic [20, 21]	The hysteretic regulator is faster than others because it does not employ a PWM. It consists of a comparator with input hysteresis that compares the output feedback voltage with a reference voltage. When the feedback voltage exceeds the reference voltage, the comparator output goes low, turning off its buck-switch MOSFET. The switch remains off until the feedback voltage falls below the reference hysteresis voltage. Then, the comparator output goes high, turning on the switch and allowing the output voltage to rise again.
Forward [22, 23]	An isolated version of the buck converter. Use of a transformer allows the forward converter to be either a step-up or step-down converter, although the most common application is step-down. The main advantages of the forward topology are its simplicity and flexibility.
Multiphase [24, 25]	The multiphase converter has multiple interleaved phases, which reduces ripple currents at the input and output. It also reduces hot spots on a p.c. board of a particular component. A two-phase buck multiphase converter reduces RMS current power dissipation in the MOSFETs and inductors by half. Multiphase cells operate at a common frequency, but are phase shifted so that conversion switching occurs at regular intervals controlled by a common control chip.
PMIC [26]	Power management integrated circuit contains multiple power supply outputs, usually two to four outputs that can range from less than 1 V to several volts. These ICs can employ internal or external power MOSFETs as part of the power supply. Usually, all the inductors are external along with output capacitors. Also, they usually incorporate overvoltage, overcurrent, short circuit, and overtemperature protection.

TABLE 4.3 ECU power supply characteristics.

Parameter	Description
Drift	Variation in DC output voltage as a function of time at constant line voltage, load, and ambient temperature.
Dynamic response	Power supply's response to a change in load power.
Hold-up time	Time during which a power supply's output voltage remains within specification following the loss of input power.
Inrush current	Peak instantaneous input current drawn by a power supply at turn-on.
Isolation	Electrical separation between the input and output of a power supply measured in volts.
line regulation	Change in value of DC output voltage resulting from a change in input voltage, expressed in ± mV or ± %.
Load regulation	Change in DC output voltage resulting from a change in load from open circuit to maximum rated output current, expressed in ± mV or ± %.
Operating temp. range	Specified operating temperature range. For example, power supply may have an operating temperature range from 20°C to +74°C.
Output noise	Occurs in short bursts of high-frequency energy. Its amplitude is variable and can depend on the load impedance and external filtering.
Periodic and random deviation (PARD)	Unwanted periodic (ripple) or aperiodic (noise) deviation of the power supply output voltage from its nominal value. Expressed in mV peak-to-peak or rms, at a specified bandwidth.
Output voltage trim	The ability to "trim" the output voltage, usually about ±10%.
Peak current	Maximum current a power supply can provide for brief periods.
Peak power	Maximum output power a power supply can provide without damage.
Ripple	Rectified and filtered switching power supply's output results in an AC component (ripple) that rides on its DC output. Ripple frequency is an integral multiple of switching frequency.
Overcurrent	Caused when the output load current is greater than specified; limited by the maximum current capability of the power supply.
Overvoltage	Occurs if the output voltage exceeds the specified DC value, which can damage the load circuits.
Soft start	Inrush current when power is first applied.
Undervoltage lockout (UVLO)	UVLO turns the supply on when it reaches a high enough input voltage and turns off the supply if the input voltage falls below a certain value.

4.4 Power Supply Selection Criteria

If you are going to make the supply, a major ECU design decision is the type of power supply system you should use [27]. Disregarding the cost, there are seven significant design considerations listed in Table 4.4

4.5 Managing Power Supply Loads

Determining ECU power requirements depends on its voltage/current ratings and the type of electronic system loads. In an EV, power is supplied from a centralized power source. To obtain this power source the HV battery voltage is applied to a DC–DC converter that produces a 12–24 VDC output. This voltage is applied to another DC–DC converter in the ECU. This could be a single output power supply or multiple outputs to supply several loads.

TABLE 4.4 ECU power supply design considerations.

Parameter	Description
Linear vs. switched mode topology	Linear power supplies always conduct current. Differences between these two configurations include size and weight, power handling capability, EMI, and regulation. Table 4.1 compares the characteristics of linear and switch-mode power supplies [28].
Dropout voltage	Typically, linear regulators require a minimum of 3 V difference between the input and output voltages. Low dropout regulators (LDOs) usually require less than 1 V differential between input and output.
Maximum output current	Regulator output current rating should be selected reasonably close to the maximum required current in the circuit. Lightly loaded regulators may have stability issues.
Voltage accuracy	This is circuit dependent. For a regulator powering digital ICs or operational amplifiers, ±5% regulation requirement is tolerable. Used as a reference voltage, accuracy is important. Accuracy is dependent on the resistance tolerance of the feedback resistors; usually ±5% tolerance is acceptable.
Power supply rejection ratio (PSRR)	Describes the capability to suppress any power supply variations in its output. Most LDOs are specified at 120 Hz. Some LDOs are powered by switch-mode power so there is a concern that harmonics of the switching frequency and any other noise sources could affect the LDO regulator input. A good figure of merit is the product of PSRR and frequency, much like gain bandwidth in an op amp.
Output noise	Power supplies may produce output noise (usually at a specific frequency) that can interfere with sensitive circuits. Spurious AC components on the output of a DC supply are called ripple and noise, or periodic and random deviation (PARD), which must be specified with a bandwidth and should be specified for both current and voltage. Because of the bandwidth consideration, PARD specifications depend on its measurement technique.
Stability	Stability controls the closed loop performance of the regulator. Poor stability means poor performance [29]. To be stable, the frequency range, or bandwidth, should correspond to a frequency where the closed loop transmission path from the input to the output drops by 3 dB (called the crossover frequency). To obtain a response converging toward a stable state, ensure that the phase is less than −180 degrees where the loop gain magnitude is 1. Therefore, tailor the loop response at the selected crossover frequency to build the necessary phase margin. The appropriate phase margin ensures that, despite external perturbations or unavoidable production spreads, changes in the loop gain will not affect system stability. Phase margin also impacts the transient response of the system.

Many of today's high-performance processing devices, such as FPGAs, ASICs, PLDs, DSPs, ADCs, and microcontrollers, require multiple voltage rails to power their own internal circuitry, such as the core, memory, and I/O. These applications demand very specific voltage rail power-up and power-down sequencing in order to guarantee reliable operation, better efficiency, and overall system health.

Sequential turning on and off of power supplies may be employed in an ECU that has multiple operating voltages. That is, power supply output voltages must be applied and removed from loads in a specific sequence. Power supply sequencing is a critical aspect to consider when designing with an FPGA or similar circuits. Failure to properly sequence the FPGA can result in excessive current drawn during startup, which can damage the FPGA. FPGAs can have from 3 to more than 10 rails, and incorrect sequencing of these rails can lead to improper device behavior. For this reason, FPGA vendors will specify the recommended power sequence for their devices.

4.6 Remote Sense

Low-voltage, high power processors may require special treatment to ensure they operate at the proper regulated load voltage. One approach is remote sense monitoring of the supply's output voltage at the load, which involves connecting plus and minus remote sense connections to the actual load voltage point, as shown in Figure 4.7. A portion of this voltage is fed back to the supply input with very little voltage drop, because the current through the two remote sense connections is very low.

Remote sensing compensates for voltage drops in the output lines (+DC OUT and −DC OUT). These voltage drops result from the resistance of the printed circuit board's (PCB) copper traces and are significant when the traces are long or not wide enough. For the best possible regulation at the load sensing traces can be relatively small, because they do not carry output current. This configuration permits the power supply to sense and compensate the voltages across the load.

4.7 Power Supply Protection

The fuse in an ECU's power supply is a part of its power management. Proper fuse selection minimizes risks and failures for an electronic system. Fuses are overcurrent devices that protect electrical and electronic devices by melting and opening a circuit to prevent excessive current from causing damage or starting a fire. For ease of maintenance, fuse replacement should be accessible externally, so it can be changed without opening the power supply enclosure.

In operation, if the current applied to the fuse exceeds its rated value, it will start to melt. Once melting begins, a gap is created, so that the current will arc across. Melting continues and the gap will grow wider until it is too wide to sustain the arc and the fuse element becomes an open circuit. Then, current ceases to flow to protect the associated circuits.

FIGURE 4.7 Remote sense capability requires +SENS and −SENS terminals of the output voltage.

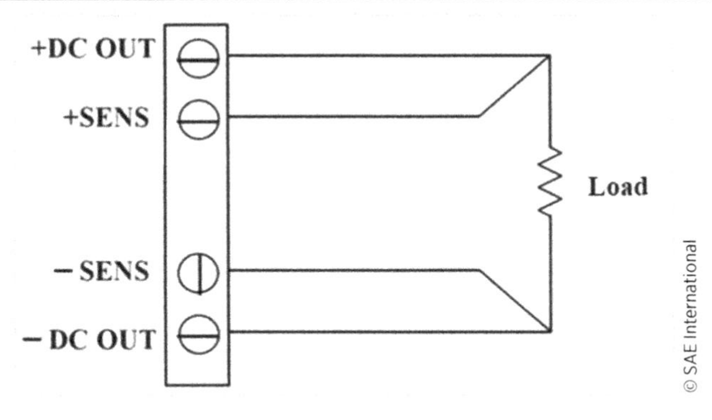

Selecting the right fuse is critical in all electronic and electrical system designs. In the event the power converter's internal circuitry can no longer withstand an overload condition, the fuse will prevent fire or further damage to a PCB, DC–DC converter, or neighboring components. Most DC–DC converters are protected from short-circuits on their outputs by either circuit-sensing current limit and/or thermal overload circuits. Fuses are required to protect against a catastrophic component failure or if a component failure creates a short-circuit on the input side of the DC–DC converter.

Chapter 9 provides more information about fuses and power malfunctions.

4.8 Optional Power Supply Features

The addition of optional features can help ECU power supply performance. One optional feature is the PMBus that provides digital control of a power supply over a specified physical bus, communications protocol, and command language. Thus, it allows the same hardware to be programmed to provide different output voltage and other performance characteristics. A conceptual diagram of PMBus-capable power supply contains a bus master and three slaves (Figure 4.8).

A typical system employing a PMBus will have a central control unit and at least one PMBus-enabled power supply attached to it. The connected power supplies are always slaves, and the central control unit is always the master. The central control unit initiates all communication on the bus, and the slave power supplies respond to the master when they are addressed.

The PMBus specification only dictates the way the central control unit and the slave power supplies communicate with each other. It does not put constraints on power supply architecture, form factor, pinout, power input, power output, or any other characteristics of the supply. The specification is also divided into two parts.

- *Part I*: Physical implementation and electrical specifications.
- *Part II*: Protocol, communication, and command language.

To be PMBus-compliant, a power supply must:

- Meet all requirements of Part I of the PMBus specification.
- Implement at least one of the PMBus commands that is not a manufacturer-specific command.
- If a PMBus command is supported, execute that command as specified in Part II of the PMBus specification.
- If a PMBus command is not supported, respond as described in the "Fault Management and Reporting" section of Part II of the PMBus specification.

Version 1.3 of PMBus was added in 2014. The major new addition to PMBus is the AVSBus, which is an interface designed to facilitate and expedite communication between an ASIC, FPGA, or processor and a POL control device on a system, for the purpose of adaptive voltage scaling. When

FIGURE 4.8 Conceptual diagram of PMBus-capable power supplies controlled from a central location.

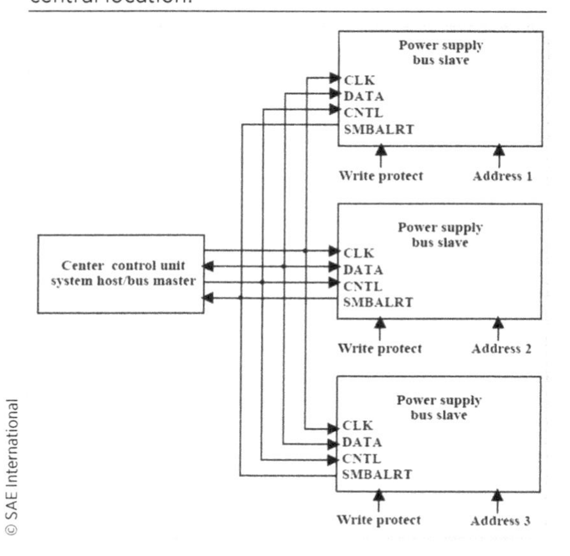

integrated with PMBus, AVSBus is available for allowing independent control and monitoring of multiple rails within one slave.

- The AVSBus is behaviorally and electrically similar to serial peripheral interface (SPI) bus without chip select lines. AVS_MData and AVS_SData are equivalent to MOSI and MISO. AVS clock is equivalent to CLK of the SPI bus. Maximum bus speed is 50 MHz.
- AVSBus is an application-specific protocol to allow a powered device such as an ASIC, FPGA, or processor to control its own voltage for power savings.
- The combination of these protocols in a slave device is an efficient and effective solution for systems containing loads that need to adapt the operating voltage.

4.9 Evaluating ECU Power Supply Performance

Determining an ECU's power requirements requires knowledge of how the selected power supply should be evaluated to see if it performs according to its specification. Evaluation of power supply performance involves four main characteristics:

- Output power
- Efficiency
- Output noise
- Reliability

4.10 Power Supply Output Power

Power supplies have a rated output power at a specific ambient temperature and input voltage. A higher ambient temperature and input voltage means you have to derate the amount of output power the supply can safely handle. Therefore, it is important to determine the ECU supply's operating temperature.

Another temperature-related function is the low-temperature "start-up" capability to turn on the supply and deliver 100% of its rated power. Additionally, the supply's output regulation, ripple and noise, and other specifications may not be its final value until the supply warms up.

4.11 Power Supply Efficiency

Many power supply characteristics impact ECU performance. One characteristic with a major effect on performance is efficiency, which relates to power supply's electrical losses and determines the amount of cooling that may be required. The ECU's other circuits can also affect efficiency, which can vary with applied voltage and output load current.

Efficiency is the ratio of total output power to input power expressed in percent. This is normally specified at full load and nominal input voltage. Therefore, power supply efficiency is the amount of the actual power delivered to its loads, divided by the input power. Figure 4.9 is a plot of efficiency for a typical buck converter.

4.12 Power Supply Output Noise

Switch-mode power supplies may produce output noise that may interfere with sensitive circuits. Output noise is usually described at a specific frequency. For example, a power supply IC's noise can be specified as $150\ \mathrm{nV}/\sqrt{\mathrm{Hz}}$ at 100 Hz.

Spurious AC components on the output of a DC supply are called ripple and noise, or periodic and random deviation (PARD), which must be specified with a bandwidth and should be specified for both current and voltage. Because of the bandwidth consideration, PARD specifications depend on its measurement technique. It is important to consider the whole signal path used to verify ripple and noise specifications. Figure 4.10 shows a measurement technique favored by Intel.

FIGURE 4.9 Typical efficiency plot for buck converter.

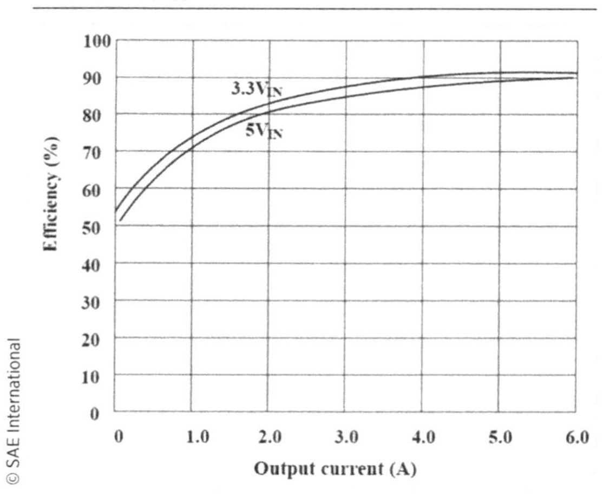

4.13 Power Supply Reliability

According to the conventional definition, reliability is defined as, "the probability that a product will perform a required function without failure under stated conditions for a stated period of time." It is based on failure rate, which is the percentage of units that will fail in a given unit of time.

Failure rate has three key phases: infant mortality, useful life, and wear-out. A high failure rate during the "infant mortality" phase is generally due to poor workmanship and shoddy components, and can be found through pre-shipment burn-in. The second, and longest, phase is "useful life," during which the supply operates properly. During this phase the failure rate is low and constant. The third phase, wear-out, is when the system is no longer usable.

A widely used approach to determining reliability is to equate it to MTBF, mean time between failures. This involves use of the original reliability prediction handbook, MIL-HDBK-217, the Military Handbook for "Reliability Prediction of Electronic Equipment." This handbook contains failure rate models for the various part types used in electronic systems, such as ICs, transistors, diodes, resistors, capacitors, relays, switches, connectors, and so on. These failure rate models are based on the best field data that could be obtained for a wide variety of parts and systems; this data is then analyzed and massaged, with many simplifying assumptions thrown in, to create usable models.

FIGURE 4.10 PARD differential mode test connection diagram suggested by Intel.

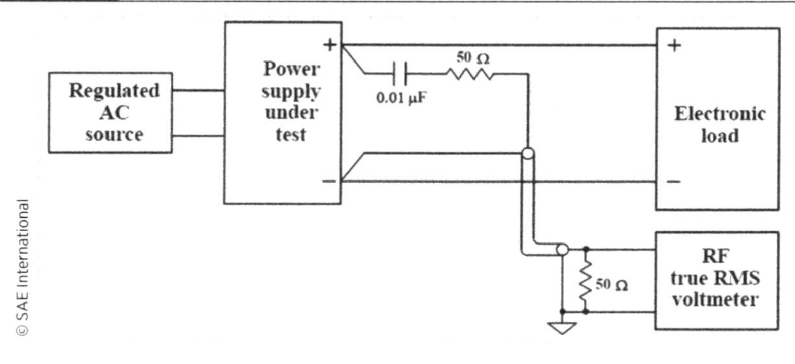

To obtain the MTBF you list all the components in the ECU power supply and assign a failure rate to each of them as given in MIL-HDBK-217. The MTBF calculation is then:

$$\mathrm{MTBF} = 1/(\text{sum of all the individual part failure rates})$$

A common misconception about MTBF is that it specifies the time (on average) when the probability of failure equals the probability of not having a failure (i.e., an availability of 50%). This is only true for certain special cases. For typical electronic products, MTBF only represents a top-level aggregate statistic and is not suitable for predicting specific time to failure.

Reliability expert Fred Schenkelberg said, "MTBF is the worst four letter acronym in our profession." Mean time between failures (or MTBF) is probably the most widely used reliability parameter. It is frequently specified as a requirement in product development specifications. Yet, when engineers are asked to explain the meaning of MTBF, only a small minority of them understands it correctly. Most people, including many engineers, think that MTBF is the same as "average life" or "useful life." This is incorrect and frequently leads to inferior design decisions.

More than just a function of failure rates, reliability is actually a function of:

- Material characteristics
- Part tolerances
- Temperature coefficients
- Design parameters
- Operating stresses
- Environmental conditions
- Customer usage
- Duty cycles
- System integration
- Interfaces between parts and subsystems
- Quality of parts and production processes

When product failure occurs and a root cause analysis is performed, it becomes evident that failures are created primarily due to errors made by design and production personnel. However, many engineers do not understand that root cause analysis should proceed to the identification of the real cause of failure. For example, when a part fails due to a very high junction temperature, high dissipated power may easily be seen as the cause of failure, while the real cause may be inferior thermal design.

ECU power supplies seldom fail due to an individual part failure, but often fail due to incorrect application and integration of its parts. Many so-called part failures can be traced to inadequate design margins (to provide robustness against manufacturing and operating variations). It should also be noted that many electronic part failures are caused by mechanical failure mechanisms. For example, a designer cannot expect reliable operation of any electronic part if the specific PCB is allowed to resonate at some frequency during operations. Failures observed during development tests, therefore, provide valuable design input, and design teams should be careful not to discard these simply as "random failures."

The book *Reliability Characterisation of Electrical and Electronic Systems*, edited by Jonathan Swingler, says the worst mistake in reliability engineering that a company can make is ignoring reliability during product development. For any product of reasonable complexity, the outcome almost certainly will be unacceptable reliability, resulting in an inferior product. In theory, a formal reliability engineering program is not a prerequisite for successful product development but, in practice, it is highly recommended.

Reliability engineering activities are often neglected during product development, resulting in a substantial increase in the risk of project failure or customer dissatisfaction. In recent years, the concept of design for reliability has been gaining popularity and is expected to continue for years to come. Reliability engineering activities should be formally integrated with other product development processes. A practical way to achieve integration is to develop a reliability program plan at the beginning of the project.

In automotive systems, semiconductors are certified by AEC-Q100, AEC-Q101, AEC-Q102, and AEC-Q104 stress tests. These tests assign grade numbers (0, 1, 2, 3) that indicate the certified operating temperature ranges. However, these tests do not specify how the device will function in actual use. Even though a device is certified for automotive use, other operational tests may be necessary. The AEC standards do not account of long-term production process stability.

4.14 Power Supply Life Tests

Predicting equipment life, such as time to failure, may be difficult because of the long operating life of many of today's products, the short time span between design and release (time-to-market), and some products are in constant use. Therefore, reliability engineers employ accelerated tests that force these products to fail more quickly than they would under normal use conditions. One way this can be achieved is by incrementally reducing the power applied to the product. You can divide accelerated life testing into two types:

- Qualitative accelerated testing that identifies failures and failure modes without attempting to predict the product's life under normal use conditions.
- Quantitative accelerated life testing predicts the life of the product, such as mean time to failure (MTTF).

Power supplies dissipate considerable power so it's probably a good idea for the ECU manufacturer to get a realistic look at the power supply's reliability using special tests. The company can send ECU's power supply or the entire ECU to a facility that performs HALT (highly accelerated life test) or ALT (accelerated life test).

HALT is the process of determining the reliability of a product by gradually varying stresses (usually temperature and/or power input) until the product fails. This is usually performed on an entire system but can be performed on individual assemblies as well.

ALT is the process of determining the reliability of a product in a short period of time by accelerating stresses (usually temperature) on the product. This is also helpful in finding dominant failure mechanisms. ALTs are usually performed on individual assemblies rather than full systems.

Results of these tests can indicate whether semiconductors or passive components will eventually fail. Performed correctly, accelerated testing can significantly reduce test times, resulting in reduced time-to-market, lower product development costs, and lower warranty costs, as well as other benefits [30].

4.15 PCB Layout Guidelines

PCB layout of any power supply is critical to its optimal performance, because poor PCB layout can disrupt the operation of an otherwise good design [31]. Even if the power

FIGURE 4.11 Simplified power supply IC with pins that are related to PCB layout.

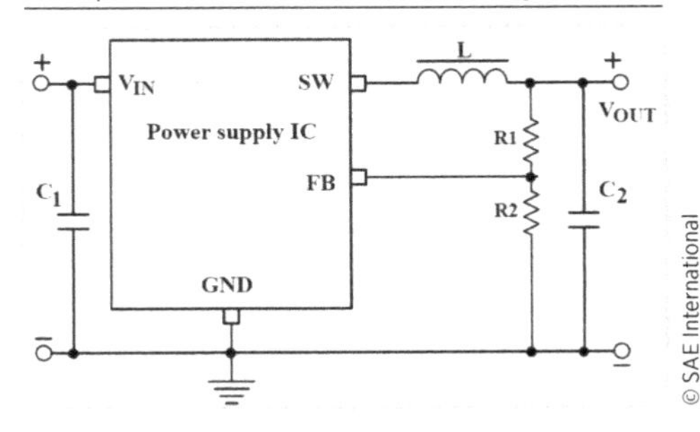

supply regulates correctly, poor PCB layout can mean the difference between a robust design and one that cannot be mass produced with consistent performance. Furthermore, the EMI performance of the power supply is also dependent on the PCB layout. In a power supply the most critical PCB feature is the loop formed by the input capacitor(s) and power ground. This loop carries large transient currents that can cause large transient voltages when reacting with the trace inductance. These unwanted transient voltages will disrupt the proper operation of the power supply. Therefore, the traces in this loop should be wide and short, and the loop area as small as possible to reduce the parasitic inductance.

Figure 4.11 shows a simplified power supply IC with pins associated with PCB layout. For best results:

1. Use a multilayer board with at least one ground plane in one of the middle layers. The ground plane acts as a noise shield and as a heat dissipation path.
2. Bypass capacitors for power input pins should be placed as close to the power supply IC as possible and have the shortest possible paths to ground.
3. Minimize parasitic inductance and resistance of all circulating high current paths by keeping board traces to all components as short and as wide as possible. Capacitor ground connections should use vias to the ground plane by way of the shortest route possible.
4. Place the feedback resistor divider as close as possible to the FB (feedback) pin of the device. Connections to FB and GND must be short and close to those pins on the device. The connection to V_{OUT} can be somewhat longer. However, this latter trace must not be routed near any noise source (such as the SW node) that can capacitively couple into the feedback path of the regulator.
5. Keep the routes that connect to the high impedance, noise-sensitive inputs FB as short as possible to reduce noise pickup.
6. Some IC packages have a thermal pad connection that must be soldered down to the PCB ground plane. This pad acts as a heat sink connection and an electrical ground connection for the regulator IC.
7. Make V_{IN}, V_{OUT}, and GND paths as wide and direct as possible, which reduces voltage drops on the input or output paths of the power supply IC and maximizes efficiency.
8. Keep the copper area connecting the SW pin to the inductor as short and wide as possible. The total area of this node should be minimized to help reduce radiated EMI.
9. To minimize the cross-sectional area of the high-frequency current loops, there should be an uninterrupted ground plane under the entire converter. This minimizes EMI and reduces the inductive drops in these loops, thereby minimizing SW pin overshoot and ringing.

4.16 Testing the System

High-performance and high-reliability ECUs typically require end-of-line testing or include automated self-testing, to assure rated performance with the supply voltage at the upper and lower limits of its regulation band. This testing is usually called "supply margining" or "voltage margining." The testing is typically accomplished by forcing the power supplies in the ECU to ±5% of their nominal operating voltage. Once the supply voltage has settled at the margined high and low voltage, you can determine ECU performance. That is, the ECU may not function properly at either the high or low voltage of its margined power supply.

Margining can be accomplished with the use of ICs intended for the purpose. These devices simplify supply margin testing and are particularly well suited for multiple power supply applications. They are available as a single power supply margining controller and a dual power supply margining controller. They provide an easy and accurate way to accomplish onboard power supply margining with a minimum of design time and board space.

Many power supply characteristics are interrelated. For example, efficiency determines the thermal and electrical losses in the system, as well as the amount of cooling required. It also impacts the physical package sizes of both the power supply and the ECU. Plus, it affects the operating temperatures of system components and the resultant system reliability. These factors contribute to the determination of the total system cost for both hardware and field support. Designing the best power supply for an ECU depends on selecting components that provide the required performance for the specific application.

4.17 AEC Standards

Specific standards and test procedures for automotive devices were established by the Automotive Electronics Council (AEC), a U.S. organization that sets qualification standards for the supply of components in the automotive electronics industry. AEC was formed in 1993 by Chrysler, Ford Motor, and Delco Electronics. AEC developed several notable specifications for automotive electronics:

- AEC Q100: "Stress Test Qualification for Integrated Circuits" (1994)
- AEC Q101: Discrete semiconductors
- AEC-Q102: Optoelectronic devices
- AEC-Q104: Multichip modules
- AEC-Q200: Passive components

The parts are assigned Grades 0–3 based on their operating temperature (Table 4.5). Qualification test sequences establish common part qualification and quality system standards for automotive electronics and is the industry standard specification that outlines the recommended qualification requirements and procedures. These tests do not cover functional testing of components, only stress tests for the components.

4.18 Industrial Regulatory Standards

There are also regulatory standards for electronic equipment that varies from country-to-country, and sometimes region-to-region, so the power supply manufacturer and the equipment manufacturer must adhere to the standards in effect where the supply is used.

Often, standards are written in a form that is difficult for the uninitiated to interpret because there are usually exemptions and exclusions that are not clear. Several different agencies may

TABLE 4.5 Operating temperature codes,

Grade	Temperature range
0	−50°C to +150°C
1	−40°C to +125°C
2	−40°C to +105°C
3	−40°C to +85°C

CHAPTER 4

be involved, so some may be specific to one country or group of countries and not others. Standard requirements vary and sometimes conflict from one jurisdiction to another.

Standards organizations include:

- ANSI: American National Standards Institute
- EC (European Community) Directives
- EN (European Norm)
- IEC (International Electrotechnical Commission)
- UL (Underwriter's Laboratory)
- CSA (Canadian Standards Association)
- ETSI (European Telecommunications Standards Institute)

Safety standards include EN60950 and UL60950, "Safety of Information Technology Equipment." Based on IEC60950, they contain requirements to prevent injury or damage due to hazards such as electric shock, energy, fire, mechanical, heat, radiation, and chemicals.

ESD (Electrostatic Discharge) standards include EN61000-4-2 that tests immunity to the effects of high voltage, low energy discharges, such as the static charge built up on operating personnel.

- EN61000-4: Tests the effects of transients
- EN61000-3-2: Limits of harmonic currents
- EN61000-4-11: Checks the effect of input voltage dips

The most commonly used international standard for emissions is CISPR. 22, "Limits and Methods for Measurement of Emissions from Information Technology Equipment (ITE)." Most of the immunity standards are contained in various sections of EN61000.

- EN61204-3: Electromagnetic compatibility (EMC) requirements
- EN61000-6-1: EMC immunity for residential, commercial, and light-industrial environments
- EN61000-6-2: Generic standards for EMC immunity in industrial environments
- EN61000-6-3: Electromagnetic compatibility (EMC) emission requirements for electrical and electronic apparatus
- EN61000-6-4: Generic EMC standards for industrial environments

Chapter 9, ECU Circuit protection, provides additional information related to regulatory standards.

References

1. Reusch, D. and Glaser, J., DC-DC Converter Handbook: A Supplement to GaN Transistors for Efficient Power Conversion (El Segundo: Power Conversion Publications, 2015), 1-185.
2. Rohm, "Basics of Linear Regulators," Rohm, December 15, 2015.
3. Day, M., "Understanding Low Dropout (LDO) Regulators," Texas Instruments, December 23, 2006.

4. Dimension Engineering, "A Beginners Guide to Switching Regulators," Dimension Engineering.
5. Rohm, "What Is a DC-DC Converter?," Rohm.
6. Lewis, L., "Switching Power Supply Tutorial," Bald Engineer, April 13, 2016.
7. Barr, M., "Introduction to Pulse Width Modulation," Embedded, August 31, 2001.
8. Selders Jr., R., "Synchronous Rectification in High-Performance Power Converter Design," Texas Instruments, September 21, 2016, 1-6.
9. Wikipedia, "Buck Converters."
10. Coates, E., "Buck Converters," Learn about Electronics, December 22, 2018.
11. Marasco, K., "How to Apply DC-DC Step-Down Regulators," Analog Dialogue, December 2011.
12. Tech Web, "Design Examples of Non-Isolated Buck Converters," Tech Web, September 7, 2017.
13. Knight, D., "Buck Converters and Their Cool Applications," All about Circuits, November 24, 2015.
14. Digi-Key, "Generating a High DC Output Voltage from a Low Input Supply," Digi-Key, January 1, 2018.
15. Wikipedia, "Boost Converter."
16. Texas Instruments, "TPIC7400-Q1 Buck and Boost Switch-Mode Regulator," Texas Instruments, December 2014.
17. Analog Devices, "LTC3115-1 40V, 2A Synchronous Buck-Boost DC-DC Converter," Analog Devices, June 11, 2015.
18. Sharp, G., "SEPIC Converter Design and Operation," Worcester Polytechnic Institute (WPI), May 11, 2014, 1-21.
19. Zhang, D., "Designing a SEPIC Converter (AN-1484)," Texas Instruments (SNVA168E), April 2013.
20. Nogawa, M., "Hysteretic-Mode Converters Demystified Part 1," Powerelectronic Technology, May 27, 2016.
21. Solis, G. and Rincon-Mora, G., "Stability Analysis & Design of Hysteretic Current-Mode Switched-Inductor Buck DC-DC Converters," Georgia Tech University, May 19, 2014, 1-4.
22. Khatri, P., "Forward Converter Circuit," Circuit Digest, September 17, 2018.
23. Wikipedia, "Forward Converter."
24. Parisi, C., "Multiphase Buck Design from Start to Finish (SLVA882)," Texas Instruments, April 25, 2017, 1-19.
25. Zhang, K., "Analysis and Design Optimization of Multiphase Converter," University of Central Florida, August 23, 2013, 1-138.
26. Wikipedia, "PMIC."
27. Davis, S., "Chapter 7: Power Supply ICs," Power Management, April 9, 2018, powerelectronics.com.
28. Rohm, "What Is the Difference between Linear and Switching Regulators," Rohm.
29. Sandler, S., Power Integrity (New York: McGraw-Hill Education, 2014), 1-335.
30. Reliasoft, "What Is Accelerated Life Test?," HBM Prenscia.
31. Texas Instruments, "AN-1149 Layout Guidelines for Switching Power Supplies," Texas Instruments, April 2013.

CHAPTER 5

Future AEC-Qualified Power Management Technology

The Automotive Electronics Council (AEC) sets U.S. qualification standards for the components supply in the automotive electronics industry. AEC was formed in 1993 by Chrysler, Ford Motor, and Delco Electronics. Since then, there have been numerous updates and modifications of the original standards.

Unfortunately, the AEC standards have not kept pace with power management technology now employed in commercial and industrial systems. These technologies include special integrated circuits (ICs) and new packaging techniques that provide smaller size modules. As more electronics are added to electric vehicles, automotive ECU designs could benefit from these new technologies, particularly with the different voltage and current requirements of the involved ICs. The device manufacturer must decide whether it is worth the time and money to test and certify the product. Additionally, the AEC should consider some devices are not covered by its current standards, such as multichip modules with discrete power semiconductors.

One of the possible AEC additions involves the stable, closed-loop performance of a power supply. To be stable, the supply's frequency range, or bandwidth, should correspond to a frequency where the closed-loop transmission path from the input to the output drops by 3 dB (called the crossover frequency). We must ensure that the phase where the loop gain magnitude is 1 is less than −180° to obtain a response converging toward a stable state Therefore, we must tailor the loop response at the selected crossover frequency to build the necessary phase margin. The appropriate phase margin ensures that, despite external perturbations or unavoidable production spreads, changes in the loop gain will not affect system stability. Phase margin also affects the transient response of the system.

The goal of this compensation is to provide the best gain and phase margins with the highest possible crossover frequency. A high crossover frequency provides a quick response to load current changes, whereas high gain at low frequencies produces fast settling of the output voltage.

Early compensation techniques provided a power supply where the designer could insert compensation components. Determining compensation component values involves an analysis of the power supply IC and its external components. After determining the required compensation, the designer modeled or measured the regulator circuit with the compensation components installed. This process usually required several iterations before obtaining the desired results.

Today, there is auto compensation that eliminates the problems associated with manual determination of power supply compensation. This relieves the designer of the need for special tools, knowledge, or experience to optimize performance. Automatic compensation sets the output characteristics, so that changes due to component tolerances, aging, temperature, input voltage, and other factors do not affect performance.

5.1 Automatic Power Supply Compensation

The Renesas (Intersil) ZL8800 provides auto compensation (Figure 5.1). It is a dual-output or dual-phase digital DC–DC controller [1]. Each output can operate independently or used together in a dual-phase configuration for high current applications. The ZL8800 supports a wide range of output voltages (0.54–5.5 V) operating from input voltages as low as 4.5 V up to 14 V.

Another automatic compensation IC from ROHM (originally designed by Powervation) that provides auto compensation is the PV4220, a dual-phase, digital

FIGURE 5.1 Renesas (Intersil) ZL8800 configured as a two-phase converter.

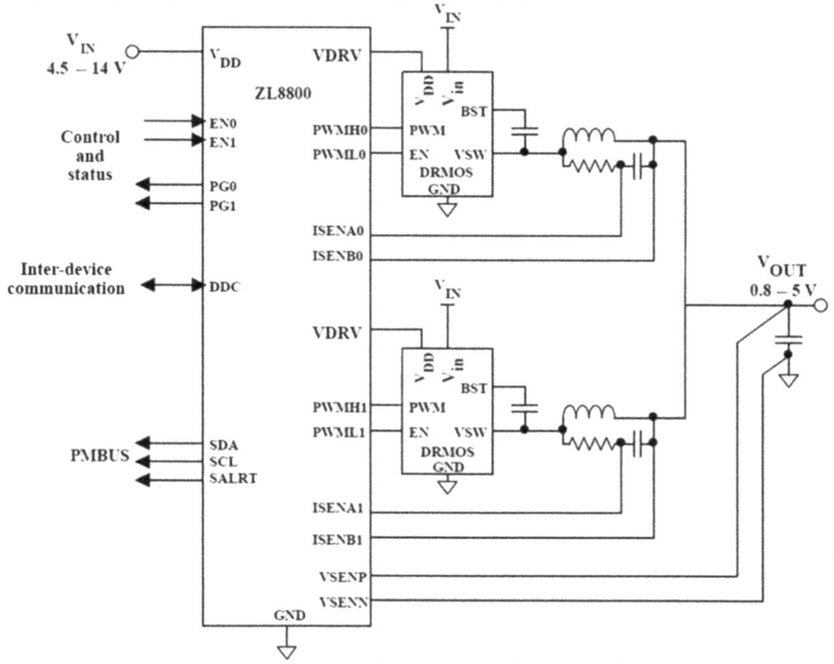

synchronous buck controller with adaptive loop compensation [2]. It has an SMBus™ Interface with PMBus™ Power System Management Protocol. Its characteristics include:

- Programmable V_{IN} range 3–100 V
- Programmable V_{OUT} range 0.45–25.6 V
- Switching frequency: 375 kHz–1 MHz
- Junction temperature range: −40°C to +125°C

5.2 Hot Swap

Another technology that could be used with ECUs is for users who want to replace a defective module without interfering with system operation [3]. They can do this by removing the existing module and inserting a new module without turning off system power, a process called "hot swap." Hot swap has been employed with systems installed in a card rack with p.c. boards. A similar approach might be applicable for ECU modules.

Figure 5.2 shows a typical hot-swap IC circuit. When inserting a plug-in module into a live chassis slot, the discharged supply bulk capacitance on the board can draw huge transient currents from the system power supplies. Therefore, the hot-swap circuit must provide some form of inrush limiting, because these currents can reach peak magnitudes ranging up to several hundred amps, particularly in high-power systems. Such large transients can damage connector pins, p.c. board etch, and plug-in and supply components. Also, current spikes can cause voltage droops on the power distribution bus, causing other boards in the system to reset. Therefore, a hot-swap control IC must provide startup current limiting, undervoltage, overvoltage, and current monitoring that prevents power supply failure.

At a hardware level, the hot-swap operation requires a reliable bus-isolation method and power management. With today's power-hungry processors, careful power ramp up and ramp down is a must in both preventing arcing on power pins and minimizing backplane voltage glitches.

FIGURE 5.2 Hot-swap control IC provides startup current-limiting, undervoltage, overvoltage, and current monitoring that prevents power supply failure.

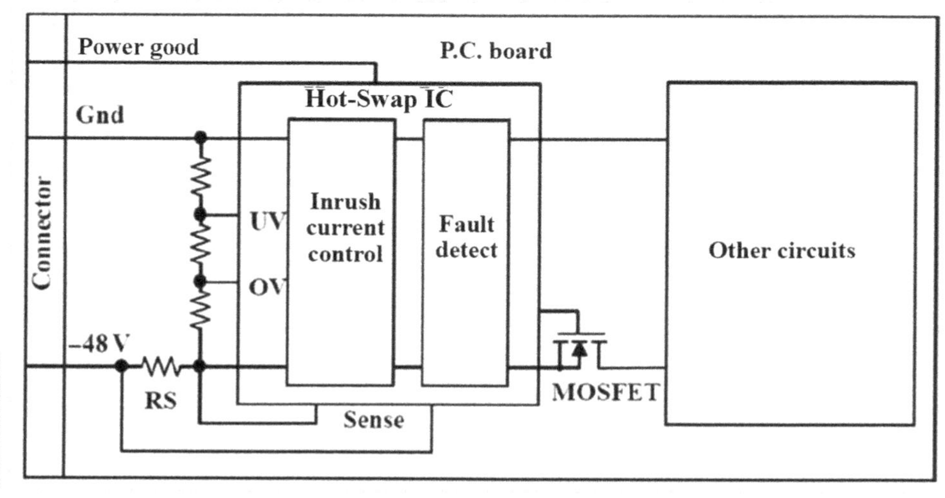

Connectors employed in these systems must also allow safe and reliable hot-swap operation. One technique is using staged pins on the backplane with different lengths. This allows events to occur in a time-sequenced manner as modules are inserted and removed. It enables the power ground and signal pins to be disconnected and then connected in an appropriate sequence that prevents glitches or arcing. After insertion, an enable signal informs the system to power up so bus-connect and software initialization can begin.

One software sequence of the extraction-insertion process starts with an interrupt signal informing the operating system of the impending event. After the operating system shuts down the board's functions, it signals the maintenance person or operator via a light-emitting diode (LED) that it is okay to remove the board. After installing a new board, the operating system automatically configures the system software. This signaling method allows the operator to install or remove boards without the extra step of reconfiguring the system.

5.3 Encapsulated Power Supplies

Encapsulated/sealed DC–DC power supplies are available in both single and multiple output versions (Figure 5.3) [4]. And there are regulated and nonregulated types. They come in SIP, DIP, SMT, and through-hole pin packages. They are usually mounted directly on a p.c. board. A typical single-output, plastic-encapsulated supply of this type is rated from 1 to 5 W. The supplies may accept nominal inputs of either 3.3, 5, 12, 15, or 24 V and produce outputs of 3.3, 5, 9, 12, 15, or 24 V. Isolation options are usually 1,000–3,000 V.

5.4 Multichip Modules

AEC-Q104 contains a set of failure mechanism-based stress tests and defines the minimum stress test-driven qualification requirements and references test conditions for qualification of multichip modules (MCMs). A single MCM consists of multiple electronic components enclosed in a single package that perform an electronic function. This document applies only to MCMs designed to be soldered directly to a printed circuit board assembly [5].

MCM types not included in the scope of this document include:

- Two assembly components or MCMs that a Tier 1/original equipment manufacturer (OEM) assembles onto a system.
- LEDs, which are covered by AEC-Q102.
- Power MCMs that may require specific considerations and qualification test procedures that are outside the scope of AEC-Q104. A power MCM consists of multiple active power devices (i.e., IGBTs, power MOSFETs, diodes) and, if necessary, additional passive devices (e.g., temperature sensors, capacitors), which are integrated on a substrate.
- Solid-state drives (SSD).
- MCMs with exterior connectors that are not soldered to a board or other assembly.

FIGURE 5.3 TRACO Power's TSR 0.5-2418 is a SIP, encapsulated, non-isolated DC–DC converter that accepts 4.75–32 VDC input and provides 1.8 VDC out at 0.5 A. Package measures 0.45 × 0.40 × 0.30 in.

FIGURE 5.4 Internal construction of the LTM8058 MicroModule (Analog Devices), an isolated DC–DC converter that provides 2.3–13 V output at 440 mA. It is housed in a 9 × 11.25 × 4.92 mm BGA package.

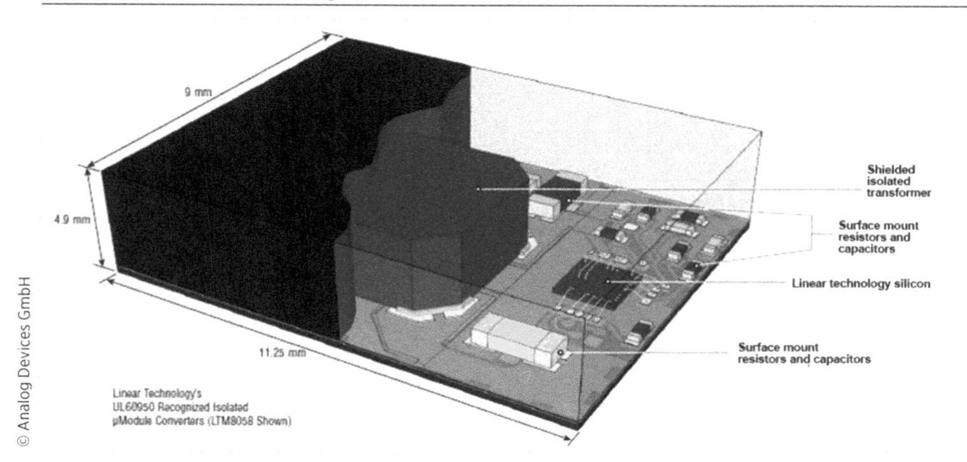

It is difficult to tell whether complete power supply circuit housed in a single package fits into the AEC-Q104 definition of an MCM, particularly its reference to including power semiconductors. Since this is an original release, updates may eventually include power semiconductors.

Power supply packages continue to shrink by using new semiconductor industry manufacturing techniques, which could be used for ECUs. For example, Analog Devices µModule® (micromodule) power supply integrates switching controllers, power FETs, inductors, and all supporting components in a conventional ball grid array (BGA) package. The Analog Devices LTM8058 µModule requires only a few input and output capacitors (Figure 5.4) [6]. The LTM®8058 is a 2 kV AC isolated µModule® DC-DC converter with LDO post regulator. Included in the package are the switching controller, power switches, transformer, LDO, and all support components. Operating over an input voltage range of 3.1–31 V, the LTM8058 supports an output voltage range of 2.5–13 V, set by a single resistor. The LTM8058 is packaged in a thermally enhanced, compact (9 mm × 11.25 mm × 4.92 mm) overmolded BGA package suitable for automated assembly by standard surface mount equipment.

A newer µModule, the LTM4700, is a non-isolated, step-down DC–DC power regulator from Analog Devices [7]. It can be configured with either a dual 50 A or single 100 A output. The technology associated with cooling and packaging the LTM4700 µModule regulator includes an overmolded package that allows insertion of the electronics as well as the cooling features. A small internal circuit board contains the electronic components (Figure 5.5). The regulator's two inductors are in a special case that helps to cool the nearby MOSFETs on the circuit board. Miniature copper piping and a cold plate aid in removing conducted heat from the device. And the package is shaped so that forced air cooling can effectively cool the µModule. Additionally, the cooling techniques were developed with an eye on cost-effective manufacturing.

FIGURE 5.5 Internal miniature circuit board on Analog Devices LTM4700 contains four Power MOSFETs, controller IC, inductors, and other components.

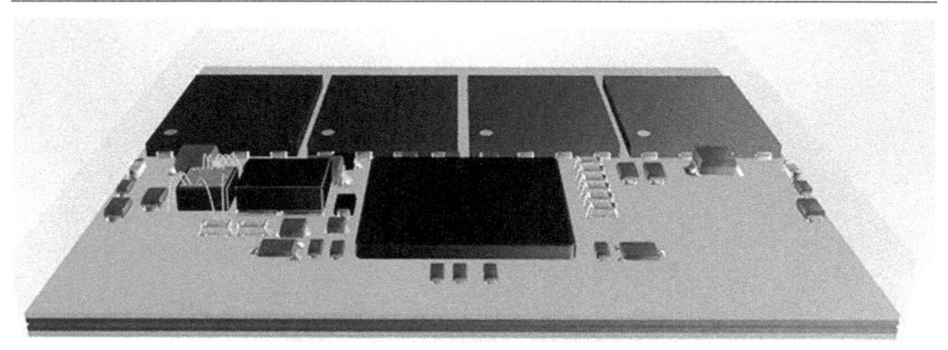

© Analog Devices GmbH

FIGURE 5.6 Texas Instruments MicroSIP 10-pin package provides a complete fixed 3.3 or 5 V output with 1 A output. Adding input and output capacitors results in a 27 mm² footprint.

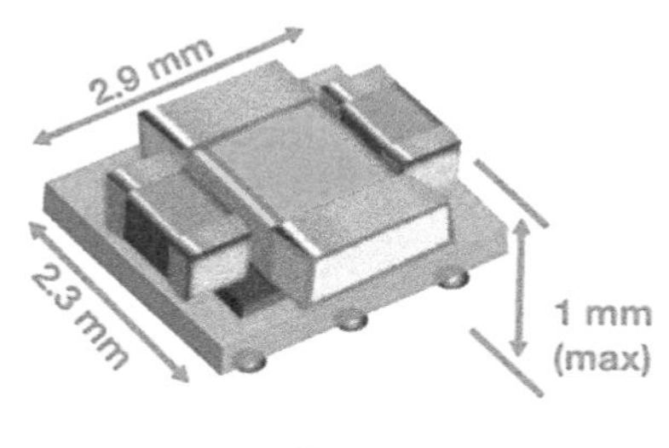

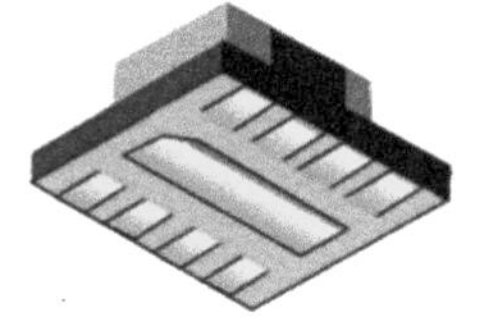

© Texas Instruments, Inc.

FIGURE 5.7 Leaded module from Texas Instruments features a flip chip or inverted die mounted on a dual lead frame.

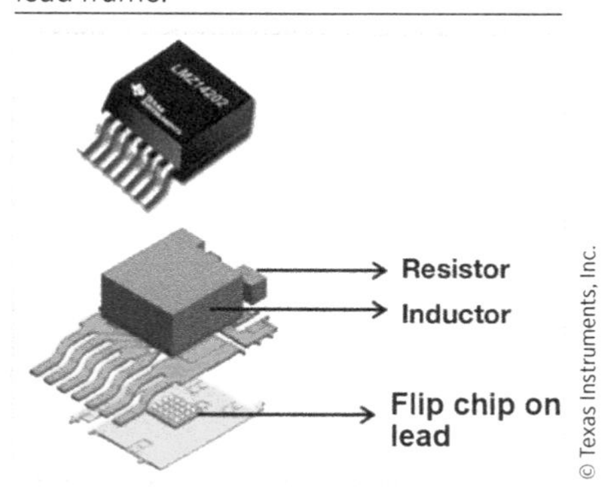

© Texas Instruments, Inc.

5.5 Nano Module

The LMZM23601 from Texas Instruments (T.I.) is a low quiescent current, step-down nano module that converts a 4–36 V DC input to a lower DC voltage with a maximum output of up to 1 A (Figure 5.6) [8, 9]. This 3.8 × 3 mm module is packaged in a MicroSiP™ that includes capacitors and inductor. It is constructed on a PCB substrate with the chip inductor and ceramic capacitors soldered on top.

This nano module requires very few external components for a complete DC–DC converter. A 3.3 or 5 V fixed output voltage option requires adding only two external components: an input and an output capacitor. An adjustable output voltage version allows the designer to set the output between 2.5 and 15 V, with the addition of two feedback resistors.

There are two current limits; a coarse high-side or peak current limit is provided to protect against faults. The high-side current limit limits the duration of the on-period of the high-side power MOSFET during a given clock cycle. A precision cycle-by-cycle valley current limit prevents excessive average output current. A new switching cycle is not initiated until the inductor current drops below the valley current limit.

T.I. also offers the LMZ14202 leaded module packaging technology that features a flip chip or inverted die mounted on a dual lead frame (Figure 5.7) [9, 10]. The dual lead frame shortens electrical paths for best-in-class EMI protection, and the copper lead frame with a thermal pad enables superior thermal performance. Passive components are stacked over the chip in the package to provide a space-saving complete power system with external leads that make mounting straightforward and provide easy access. These modules include ruggedized options for applications in harsh environments.

The LMZ14202 requires three external resistors and four external capacitors. It operates from a 4.5 to 42 V input and can supply up to a 12 W output at 2 A output. The output is adjustable for 0.8-6 V. It includes an integrated shielded inductor in its 9.85 × 10.16 mm package that has seven leads that are 3 mm long.

References

1. Renesas, "ZL8800 Dual Channel/Dual Phase PMBus ChargeMode Control DC/DC Digital Controller," Renesas Data Sheet, July 18, 2018, 1-88.
2. Rohm Semiconductor, "PV4022 Dual-Phase Digital DC/DC Buck Controller with Auto-Control®, & SMBus™/PMBusTM," Rohm Semiconductor Data Sheet, 1-3.
3. Hegarty, T., "TVS Clamping in Hot-Swap Circuits," Power Electronics Technology 37, no. 10 (July 27, 2011).
4. TRACO POWER, "TSR 0.5 Series, 0.5 A Switching Regulator," TRACO POWER Data Sheet, September 10, 2018, 1-3.
5. AEC, "Failure Mechanism Based Stress Test Qualification for Multi-Chip Modules (MCM)," AEC.
6. Analog Devices, "LTM8058 3.1VIN to 31VIN Isolated uModule DC/DC Converter with LDO Post Regulator," Analog Devices Data Sheet, July 14, 2017, 1-20.
7. Analog Devices, "LTM4700 Dual 50A or Single 100A iModule Regulator with Digital Power System Management," Analog Devices Data Sheet, October 9, 2018, 1-126.
8. Texas Instruments, "LMZM23601 36-V, 1-A Step-Down DC/DC Power Module in 3.8-mm × 3-mm Package," Texas Instruments Data Sheet, December 27, 2018, 1-45.
9. Chaudry, U., DeVries, C., Kummerl, S., and Chong, H.L., "Powerful Solutions Come in Small Packages (sszy021)," Texas Instruments.
10. Texas Instruments, "LMZ14202 SIMPLE SWITCHER® 6V to 42V, 2A Power Module in Leaded SMT-TO Package," Texas Instruments Data Sheet, January 9, 2019, 1-31.

CHAPTER 6

EV Power Semiconductors

Power semiconductors control systems that handle power levels up to kilowatts. They also are an important component is power supplies. Typically, power semiconductors act as electronic switches that control the turn-on and turn-off of high-power, electronic loads.

Figure 6.1 is a representation of a power semiconductor switch. The power semiconductor switch applies power to a load when a control signal tells it to do so [1]. The control signal also tells it to turn off. Ideally, the power semiconductor switch should turn on and off in zero time. It should have an infinite impedance when turned off, so that zero current flows to the load. It should also have zero impedance when turned on, so that the on-state voltage drop is 0 V. Another idealistic characteristic would be that the switch input consumes zero power when the control signal is applied. However, these idealistic characteristics are unachievable with the present state of the art.

In the real world, actual power semiconductor switches do not meet the ideal switching characteristics. For example, Figure 6.2(a) shows a control signal applied to a power semiconductor switch whose ideal output exhibits zero transition time when turning on and off (Figure 6.2(b)). When the transistor is off (not conducting current) power dissipation is very low because current is very low.

An actual power semiconductor switch exhibits some delay when turning on and off, as shown in Figure 6.2(c). Therefore, some power dissipation occurs when the switch goes through the linear region from on to off. This means that the most power dissipation depends on the time spent going from the off to on and vice versa, that is, going through the linear region. Thus, the faster the device goes through the linear region, the lower the power dissipation and losses.

When the power semiconductor switch is on (conducting maximum current) power dissipation is low because its on-resistance is low. When turned on, the switch voltage is 0 V plus the device current, I, times the power semiconductor's on-resistance, R_{ON}.

FIGURE 6.1 Power semiconductor switch turns power to the load on and off.

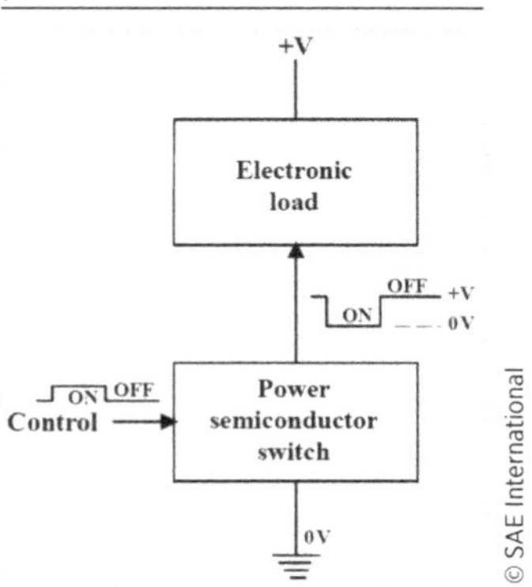

FIGURE 6.2 Waveforms for the power semiconductor switch showing the control input (a), the ideal output (b), and the actual output (c).

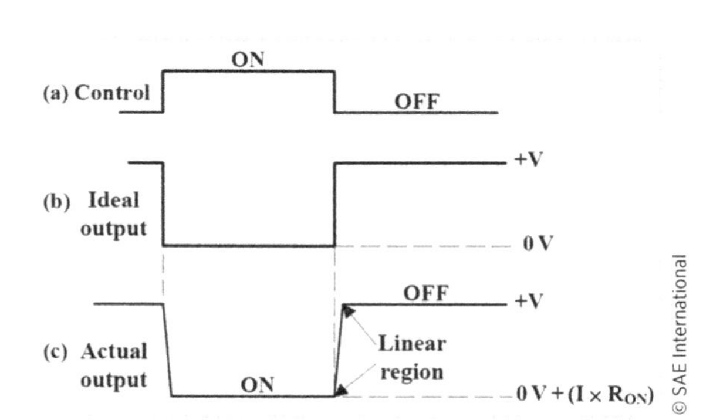

6.1 Power MOSFETs

Today, the most widely used power semiconductor switch is the power MOSFET (metal-oxide semiconductor field effect transistor), a three-terminal silicon device consisting of a gate, drain, and source [2–5]. Applying a signal to its gate turns the power MOSFET switch on, and when that signal is removed the power MOSFET turns off. There are two types of MOSFETs: n-channel versions require a positive gate turn-on voltage and p-channel devices require a negative gate turn-on voltage. Their current conduction capabilities are up to several tens of amperes, with typical breakdown voltage ratings of up to 1,000 V.

The MOSFET turns on when its gate-to-source voltage is above a specific threshold. Typical gate thresholds range from 1 to 4 V. For an n-channel MOSFET, a positive bias greater than the gate-to-source threshold voltage is applied to the gate, and a current flows between source and drain. For gate voltages less than the threshold, the device remains in the off state. P-channel MOSFETs use a negative gate drive signal to turn a MOSFET on. MOSFET gate drivers must supply enough drive to ensure that the power switch turns on properly. Some gate drivers also have protection circuits to prevent failure of the power semiconductor switch and its load [3, 4]. Figure 6.3 shows details of the power MOSFET inputs, including parasitics.

FIGURE 6.3 Typical N-Channel MOSFET's parasitic capacitances includes CGD (gate-to-drain), CDS (drain-to-source), and CGS (gate-to-source).

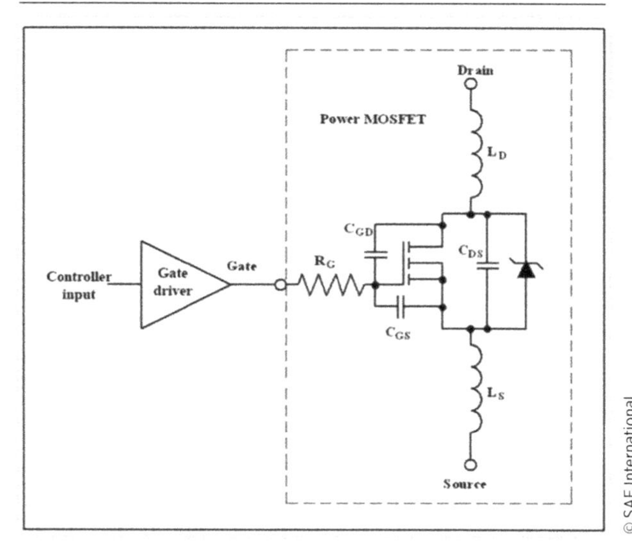

Industry estimates are that the average internal combustion automobile now uses about 60 power MOSFET semiconductors to control electronic actuators, as well as switching lamps, motors, and injectors. The move to electric vehicles will increase the number of power MOSFETs per automobile. Therefore, power MOSFETS are an important automotive component. For automotive applications, power MOSFETs must be qualified to AEC-Q101 [1, 2].

Another major application for power MOSFETs is in switch-mode power supplies. In some applications the MOSFET is integrated within the power supply IC. In other applications the power MOSFETs are external and connect with the power supply IC.

6.2 MOSFETs and BJT Comparison

Before MOSFETs were used in automotive applications bipolar junction transistors (BJTs) were used. Comparing the two devices:

- Power MOSFETs of similar size and voltage rating can operate at higher frequencies than BJTs.
- Typical power MOSFET rise and fall times are two orders of magnitude faster than BJTs of similar voltage rating and active area.
- BJTs are limited to frequencies of less than 100 kHz, whereas power MOSFETs can operate up to several MHz before switching losses become unacceptably high.
- Power MOSFETs are voltage-controlled devices with simple drive circuitry requirements.
- Power BJTs are current-controlled devices requiring large base drive currents to keep the device in the ON state.

The forward voltage drop of a BJT decreases with increasing temperature potentially leading to destruction. This is important when several devices are paralleled to reduce forward voltage drop. Power MOSFETs can be paralleled easily, because their forward voltage increases with temperature so there is an even distribution of current among all components. MOSFETs can withstand simultaneous application of high current and high voltage without undergoing destructive failure due to a process called second breakdown. However, at high breakdown >200 V, the on-state voltage drop of the power MOSFET becomes higher than that of a similar sized bipolar device with similar voltage rating, making it more attractive to use the bipolar power transistor despite exhibiting worse high-frequency performance.

Bipolar transistors are rated for maximum current under continuous and pulsed conditions; exceeding these ratings can cause device failure. MOSFET behave like a resistor when they turn on, so the maximum voltage drop or heat generated determines the maximum current. Turning MOSFET current on and off at high speeds reduces the average power as well as generated heat, thereby increasing the maximum allowable current.

Power MOSFET characteristics include several parameters critical to their performance, as listed in Table 6.1.

6.3 MOSFET Fabrication Technologies

MOSFETs are classified as planar and superjunction types, according to the manufacturing processes used. A planar structure has the drawback of relatively high on-resistance at higher operating voltage. A superjunction structure has multiple vertical pn junctions, resulting in a lower on-resistance and reduced gate charge, which allows more efficient switching at higher operating voltage.

There is also a trench gate MOSFET whose goal is to make the device conduct current vertically from one surface to the other to achieve a high cell density and low on-resistance. Although trench technology allows a higher cell density, it is more difficult to manufacture than the planar device. Process refinements have yielded devices with steadily increasing density and lower on-resistance. Trench MOSFETs have achieved on-resistance less than 1 mΩ for a 25 mm^2 silicon die, exclusive of lead resistance.

TABLE 6.1 Power MOSFET characteristics [5].

Parameter	Description
Blocking Voltage	Maximum voltage that can be applied to a MOSFET, including the applied voltage plus any inductively induced voltage.
Maximum Single Pulse Avalanche Energy	Determines how much energy the MOSFET can withstand under avalanche conditions. Avalanche occurs if the maximum drain-to-source voltage is exceeded and current rushes through the device.
On-Resistance	Determines the power loss and heating of the power semiconductor. The lower the on-resistance the lower the device power loss and the cooler it will operate. Low on-resistance drastically reduces heat-sinking requirements in many applications, which lowers parts count and assembly costs. Usually described as $R_{DS(ON)}$, or resistance from drain-to-source when turned on.
Maximum Junction Temperature	A function of the electrical characteristics of the device itself, as well as the package in which it is housed. Package thermal properties determine its ability to extract heat from the die. The junction-to-ambient and junction-to-case thermal resistance is a measure of the MOSFET's ability to extract heat.
Drain Current	MOSFET output current for driving a specific load. This value can be limited by the MOSFET's package. When operated in the pulsed mode, the MOSFET's drain current can be several times its continuous rating.
Safe Operating Area	A function of the voltage and current applied to the device. A curve in power transistor data sheets that defines the allowable combination of voltage and current, the device can safely handle.
Total Gate Charge	The charge on the gate as determined by its gate-to-source capacitance. The lower the gate charge, the easier it is to drive the MOSFET. Total gate charge, QG, affects the highest reliable switching frequency of the MOSFET.
Figure of Merit	Relates to the tradeoff between on-resistance and gate charge. The product of the on-resistance and gate charge is a figure of merit (FOM) that compares different power MOSFETs for use in high-frequency applications.
Threshold Voltage	Minimum gate source electrode bias required to form a conducting channel between the source and the drain regions.
Power Loss	Conduction losses, ruggedness, and avalanche capability are important features. Conduction losses are determined by the product of operating current and on-resistance of the power MOSFET.
Maximum Allowable Power Dissipation	This is the power dissipation that raises the maximum allowable junction temperature, when the case temperature is held at 25°C, it is normally 150°C or 175°C.
Body Diode Forward Voltage	The guaranteed maximum forward drop of the body-drain diode at a specified value of source current.
Thermal Resistance, Junction-To-Case	A typical surface mount package can have a thermal resistance of 30–50°C/W, whereas a typical TO-220 device can be 2°C/W or less.
Maximum dv/dt	Maximum rate of rise of source-drain voltage allowed. Under certain conditions a catastrophic failure may occur.
Electrostatic Discharge	ESD is the static charge accumulated by a person handling an MOSFET semiconductor is often enough to destroy the part.
Unclamped Inductive Switching	If current through an inductance is turned off quickly, the resulting magnetic field induces a counter electromagnetic force that can build up surprisingly high potentials across the MOSFET. If this induced potential exceeds the rated voltage breakdown it can cause a catastrophic failure.

Proprietary MOSFET processing techniques have been developed that yield variations in on-resistance and gate charge that affect device efficiency and switching frequency.

STMicroelectronics claims its MDmesh power MOSFETs are optimized for high-power power factor correction (PFC) and PWM topologies in hard switching applications power supplies for EV/HEV [6]. They provide:

- Extremely low on-resistance) for increased efficiency and more compact designs
- Optimized tradeoff between on-resistance and capacitance profiles for increased performance in high-power applications

STMicroelectronics also produces the STripFET™ F7 series MOSFETs that feature an enhanced trench gate structure that lowers device on-state resistance while also reducing internal capacitances and gate charge for faster and more efficient switching [7]. Its characteristics include:

- Low on-resistance
- Low FOM for increased system efficiency and more compact designs
- Optimized parasitic capacitance for improved EMI immunity
- High avalanche ruggedness

Infineon CoolMOS™ superjunction power MOSFETs have gone through several generations of products and are available with operating voltages up to 900 V [8]. Its characteristics include:

- Reduced on-resistance area product by a factor of five for 600 V transistors, which redefines the dependence of on-resistance on breakdown voltage.
- Due to chip shrink and novel internal structure, the technology shows a very small input capacitance as well as a strongly nonlinear output capacitance.
- Lower gate charge facilitates and reduces the cost of controllability, and the smaller feedback capacitance reduces dynamic losses.

Infineon also produces the OptiMOS family of trench and planar MOSFETs that handle up to 300 V.

Texas Instruments introduced a third-generation power MOSFET, NexFET™ technology [9]. It offers an on-resistance competitive with the TrenchFET, which reduces the input and Miller capacitances significantly. Low capacitances mean low input gate charge and short voltage transients during switching. This new generation MOSFET reduces switching losses in switch-mode power supply applications and enables operation at high switching frequencies. This technology is most advantageous at 30 V and below.

FemtoFET™ N-Channel MOSFET transistors from Texas Instruments are said to be the smallest, low on-resistance Power MOSFETs available today [10]. The FemtoFET is housed in a land grid array (LGA) package, which is a silicon chip-scale package with metal pads instead of solder balls. The FemtoFET is ideal for applications where saving board space and extending battery life are required.

Among the companies that have announced power MOSFETs qualified for AEC-Q101 are:

- ON Semiconductor/Fairchild
- Diodes Incorporated
- STMicroelectronics
- Vishay/Siliconix
- Infineon
- Nexperia
- Toshiba
- Littelfuse/IXYS
- Rohm

6.4 MOSFET Packages

MOSFETs are available in small outline IC (SOIC) packages for applications where space is at a premium. Larger through-hole TO-220, TO-247, and the surface mountable D^2PAK or SMD-220 are also available. Devices with breakdown voltage ratings of 55–60 V and gate threshold voltages of 2–3 V are used mainly in through-hole packages such as TO-220, TO-247, or the surface-mounted D^2PAK (SMD-220). These through-hole packages have very low thermal resistance. Despite their higher thermal resistances, more surface-mounted SOIC packages are finding their way into applications due to the continuous reduction in on-resistance. SOIC packages save space and simplify system assembly. The newest generation of power MOSFETs use chip-scale and ball grid array packages for low voltage power MOSFETs.

The Infineon (International Rectifier) DirectFET2 MOSFETs is an AEC-Q101 qualified surface-mounted power MOSFET technology (Figure 6.4) [11, 12]. The DirectFET2 metal encased MOSFET family offers matching 20 and 30 V synchronous buck converter MOSFET chipsets, followed by the addition at 30 V targeted for high-frequency operation. The DirectFET2 MOSFET family is also available in three different can sizes. Connections to the gate and source are in solder pads on the bottom of the device. The drain connects to tabs on both ends of the metal enclosure. Solder reflow is used to electrically connect the solder pads and drains to a p.c. board. Table 6.2 lists the characteristics of the AUIRF7736M2TR DirectFET2.

Vishay's PolarPAK (Figure 6.5) is an AEC-Q101-qualified, thermally enhanced package that facilitates MOSFET heat removal from an exposed top metal lead frame connected to a drain surface in addition to a source lead frame connected to a PCB [13]. PolarPAK

FIGURE 6.4 Internal construction of the Infineon (International Rectifier) DirectFET2®, an AEC-Q101 qualified surface mount semiconductor for board-mounted power applications.

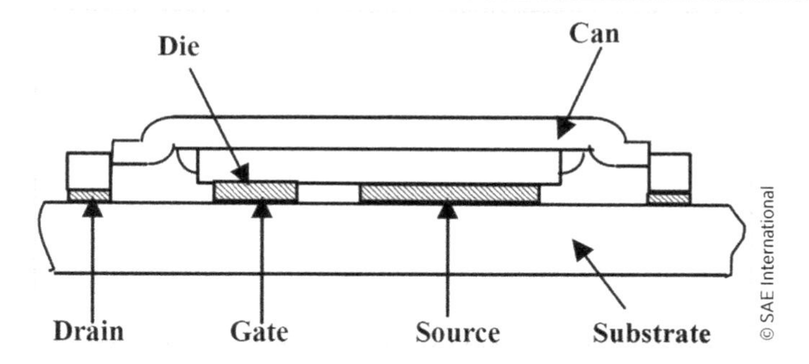

TABLE 6.2 Characteristics of the Infineon AUIRF7736M2TR.

AUIRF7736M2TR DirectFET2 at 25°C	
Parameter	**Value**
Length (max)	6.35 mm
Width (max)	5.05 mm
Height (max)	0.74 mm
V_{DSS} (max)	40 V
I_D (max)	108 A
$R_{DS(ON)}$ (max)	3.0 m Ω
Q_G (typ)	72 nC

was specifically designed for easy handling and mounting onto the PCB with high-speed assembly equipment and thus to enable high assembly yields in mass volume production. PolarPAK power MOSFETs have the same footprint dimensions of the standard SO-8, dissipate 1°C/W from their top surface and 1°C/W from their bottom surface. This provides a dual heat dissipation path that gives the devices twice the current density of the standard SO-8. With its improved junction-to-ambient thermal impedance, a PolarPAK power MOSFET can either handle more power or operate with a lower junction temperature. A lower junction temperature means a lower $R_{DS(ON)}$, which means higher efficiency.

FIGURE 6.5 The Vishay Polar PAK™ package increases the power handling capability of power MOSFETs, while keeping a PCB landing pattern no bigger in area than that of a standard SO-8 orPowerPAK® SO-8.

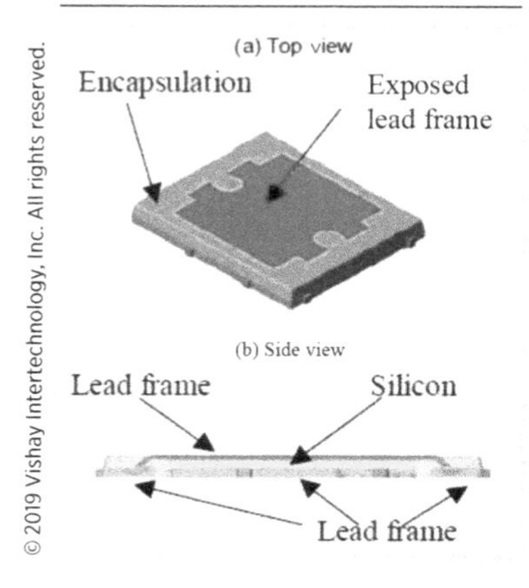

6.5 DrMOS

Intel's DrMOS specification identified a multichip module consisting of a gate driver and power MOSFET [14]. A major advantage of using this module (Figure 6.6) is that the individual MOSFET's performance characteristics can be optimized, whereas monolithic MOSFETs produce higher on-resistance. Although the component cost of a multichip module may be higher than a monolithic part, the designer should view the cost from a system viewpoint. That is, space is saved, potential noise problems are minimized, and fewer components reduce production time and cost. Here, a multichip approach would be superior to use of a monolithic part.

Unlike discrete solutions whose parasitic elements, combined with board layout, significantly reduce system efficiency, the DrMOS module is designed to both thermally and electrically minimize parasitic effects and improve overall system efficiency. In operation, the high-side MOSFET is optimized for fast switching while the low-side device is optimized for low on-resistance. This arrangement ideally accommodates the low-duty-cycle switching requirements needed to convert the 12 V bus to supply the processor core with 1.0–1.2 V at up to 30 A.

FIGURE 6.6 DrMOS is a multichip module that contains two MOSFETs and the associated drive circuits.

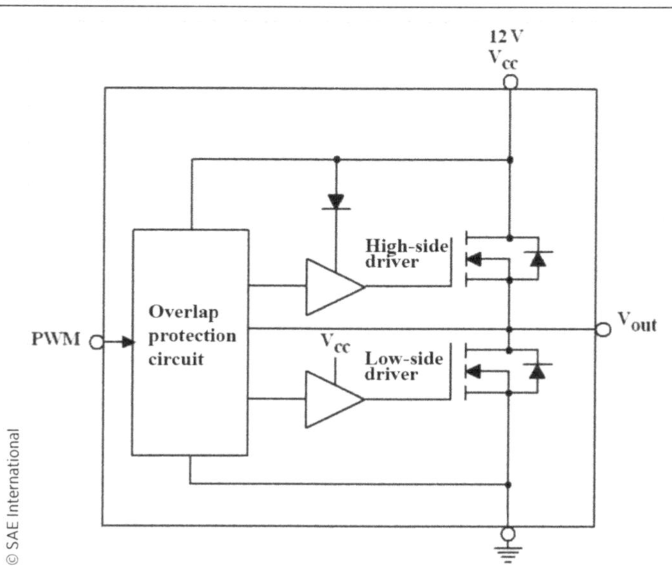

CHAPTER 6

6.6 Power MOSFETs for Automotive Applications

An important feature of these new MOSFETs is their moisture sensitivity level, or MSL. This relates to the packaging and handling precautions for semiconductors and is an electronic standard for the time the device can be exposed to ambient room conditions of approximately 30°C/60%RH. The reason this is important is that thin fine-pitch devices could be damaged during surface mount technology (SMT) reflow manufacture when moisture trapped inside the component expands. Trapped moisture can damage a semiconductor. In extreme cases, cracks will extend to the component surface.

According to IPC/JEDEC's J-STD-20: "Moisture/Reflow Sensitivity Classification for Plastic Integrated Circuit (IC) SMTs," there are eight levels of moisture sensitivity [15]. Originally, MOSFETs were rated at MSL 3, which allows 168 h of moisture testing. MSL 1 allows an unlimited time for the moisture test.

Today, there are new applications for power MOSFETs. This includes electric power steering (EPS), drive train, and power train systems. Besides meeting the AEC-Q101 standard, they must meet the cost constraints imposed by automotive manufacturers.

For EPS, an electric motor driven by a power MOSFET provides steering assist to the driver of a vehicle. A typical system employs sensors that detect motion and torque of the steering column, and an ECU module controls system performance. Software allows varying amounts of assistance to be applied depending on driving conditions.

EPS systems have a fuel efficiency advantage over conventional hydraulic power steering. The electrical approach eliminates the belt-driven hydraulic pump constantly running, whether assistance is required or not. Another major advantage is the elimination of a belt-driven engine accessory and several high-pressure hydraulic hoses between the hydraulic pump, mounted on the engine, and the steering gear mounted on the chassis. This simplifies manufacturing and maintenance.

Power MOSFETs have played a major role in making internal combustion engine (ICE) vehicles more reliable and will do the same for EVs. Among the traditional mechanical components that have been eliminated are shafts, pumps, hoses, fluids, coolers, and so on, which reduces the weight of the vehicle and improves fuel efficiency. Safety improvement is another feature of electronic controls that provide more automated functions that cannot be achieved by mechanical techniques. Compared with mechanical systems, the electronics trend also allows easier modification or upgrade of automotive systems.

6.7 Power Semiconductor Reliability

Excessive operating voltage can cause power semiconductor failures because the devices may have small spacing between their internal elements. An even worse condition for a power semiconductor is to have high voltage and high current present simultaneously. A few nanoseconds at an excessive voltage or excessive current can cause a failure. Most power semiconductor data sheets specify the maximum voltage that can be applied under all conditions. The military has shown very clearly that operating semiconductors at 20% below their voltage rating provides a substantial improvement in their reliability.

Another common killer of power semiconductors is heat. Not only does high temperature destroy devices, but even operation at elevated, nondestructive temperatures can degrade useful life. Data sheets specify a maximum junction temperature, which is typically between 100°C and 200°C for silicon. Most power transistors have a maximum junction rating of 125°C–150°C, the safe operating temperature is much lower. Grade 1 AEC-Q101-qualified power semiconductors are listed as −40°C to 125°C.

6.8 Insulated Gate Bipolar Transistor (IGBT)

An insulated gate bipolar transistor (IGBT) is a three-terminal power semiconductor device primarily used as an electronic switch that combines high efficiency and relatively fast switching. The IGBT provides the simple gate drive characteristics of MOSFETs with the high current and low-saturation voltage capability of bipolar transistors. It combines an insulated gate FET for the control input and a bipolar power transistor as a switch in a single device [16, 17].

The IGBT is used in medium-to-high-power applications like switch-mode power supplies and traction motor control. Large IGBT modules typically consist of many devices in parallel and can have very high current-handling capabilities in the order of hundreds of amperes with blocking voltages of 6,000 V. These IGBTs can control loads of hundreds of kilowatts. It is equally suitable in resonant-mode power supply circuits [18, 19].

The main advantages of IGBT over a power MOSFET and a BJT are:

1. Low on-state voltage drops due to conductivity modulation and on-state current density. So smaller chip size is possible, reducing cost.
2. Low driving power and a simple drive circuit due to the input MOS gate structure, which allows relatively easy control compared with current-controlled devices in high-voltage and high current applications.
3. Compared with a bipolar transistor, the IGBT has better current conduction capability as well as forward and reverse blocking capabilities.
4. Other IGBT characteristics:
   - Slower switching speed compared with a power MOSFET, but better than a BJT.
   - Possibility of latch-up due to the internal PNPN thyristor structure.

A simple equivalent circuit model of an IGBT is shown in Figure 6.7. It contains MOSFET, JFET, NPN, and PNP transistors. The collector of the PNP is connected to the base of the NPN, and the collector of the NPN is connected to the base of the PNP through the JFET. The NPN and PNP transistors represent the parasitic thyristor that constitutes a regenerative feedback loop. Resistor RB represents the shorting of the base-emitter of the NPN transistor to ensure that the thyristor does not latch-up, which will lead to the IGBT latch-up. The JFET represents the constriction of current between any two neighboring IGBT cells. It supports most of the voltage and allows the MOSFET to be a low voltage type and consequently have a low on-resistance value.

In general, high-voltage, high current, and low-switching frequencies favor the IGBT, while low voltage, low current, and high switching frequencies are the domain of the MOSFET.

FIGURE 6.7 Simplified IGBT equivalent circuit.

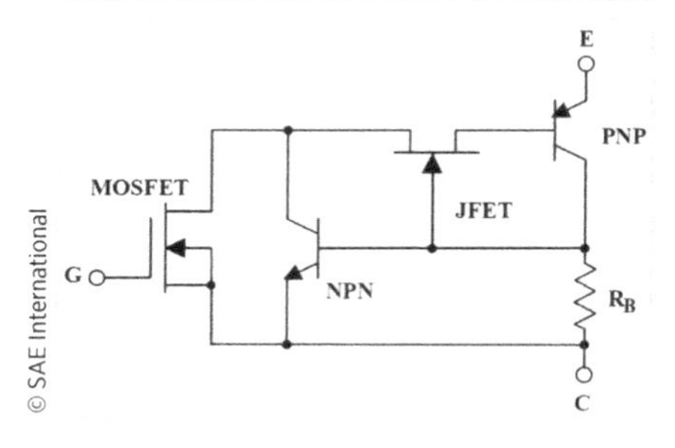

There two types of IGBTs: non-punch-through (NPT) and punch-through (PT). The PT type has an extra buffer layer that performs two functions:

- Avoids failure by PT action because the depletion region expansion at applied high voltage is restricted by this layer.
- Reduces the tail current during turn-off and shortens the fall time of the IGBT.

NPT IGBTs have equal forward and reverse breakdown voltage, so they are suitable for AC applications. PT IGBTs have less reverse breakdown voltage than the forward breakdown voltage, so they are applicable for DC circuits where devices are not required to support voltage in the reverse direction.

The IGBT has a much lower "on-state" resistance than an equivalent MOSFET. This means that the I^2R drop across the bipolar output structure for a given switching current is lower. The forward blocking operation of the IGBT transistor is identical to a power MOSFET.

When used as static controlled switch, the IGBT has voltage and current ratings similar to that of the bipolar transistor. However, the presence of an insulated gate in an IGBT makes it a lot simpler to drive than the BJT as it requires much less drive power.

An IGBT is turned "ON" or "OFF" by activating and deactivating its gate terminal. Applying a positive input voltage signal across the gate and the emitter will keep the device in its "ON" state, while making the input gate signal zero or slightly negative will cause it to turn "OFF" in much the same way as a bipolar transistor or MOSFET. Another advantage of the IGBT is that it has a much lower on-state channel resistance than a standard MOSFET.

The IGBT is a voltage-controlled device, so it only requires a small voltage on the gate to maintain conduction through the device unlike BJT's that require that the base current is continuously supplied in a sufficient quantity to maintain saturation.

Also, the IGBT is a unidirectional device, meaning it can only switch current in the "forward direction," that is from collector to emitter unlike MOSFETs that have bidirectional current-switching capabilities (controlled in the forward direction and uncontrolled in the reverse direction).

The principle of operation and gate drive circuits for the IGBT are similar to that of the N-channel power MOSFET. The basic difference is that the resistance offered by the main conducting channel when current flows through the device in its "ON" state is much smaller in the IGBT. Because of this, the current ratings are much higher than an equivalent power MOSFET.

The main advantages of using the IGBT over other types of transistors are:

- High-voltage capability
- Low on-resistance
- Ease of drive
- Relatively fast switching speeds
- Zero gate drive current

These characteristics make the IGBT a good choice for moderate speed, high-voltage applications such as in PWM, variable speed control, DC–AC inverters, and frequency converter applications operating in the hundreds of kilohertz range.

Another advantage of the IGBT transistor is the simplicity by which it can be driven on by applying a positive gate voltage or switched off by making the gate signal zero or slightly negative, allowing it to be used in a variety of switching applications.

With its lower on-state resistance and conduction losses, as well as its ability to

FIGURE 6.8 Output characteristics of the 160 A, 650 V ON Semiconductor FGY160T65SPD at 25°C.

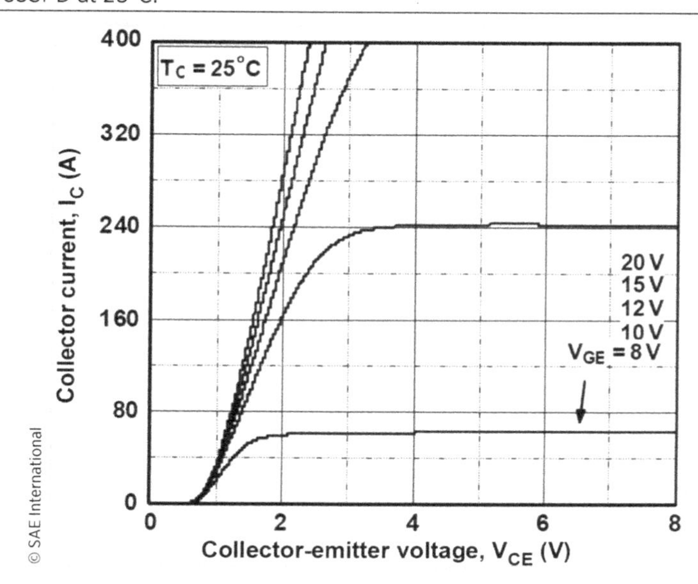

switch high voltages at high frequencies without damage, the IGBT is ideal for driving inductive loads such as coil windings, electromagnets, and DC motors. Figure 6.8 shows the relationship between the IGBT's collector voltage and current.

6.9 Wide Bandgap Semiconductors

A key semiconductor characteristic is bandgap that indicates the amount of energy required to jolt an electron into a conducting state [20]. A wide bandgap (WBG) enables higher power, higher switching speed transistors. WBG devices include gallium nitride (GaN) and silicon carbide (SiC) that are listed in Table 6.3 along with other semiconductors.

WBG benefits include:

- Elimination of up to 90% of the power losses that occur during power conversion.
- Up to 10 times higher switching frequencies than Si-based devices.
- Operation at higher maximum temperature than Si-based devices.
- Systems with reduced lifecycle energy use.

TABLE 6.3 Semiconductor bandgap comparison.

Material	Chemical symbol	Bandgap energy (eV)
Germanium	Ge	0.7
Silicon	Si	1.1
Gallium Arsenide	GaAs	1.4
Silicon Carbide	SiC	3.3
Gallium Nitride	GaN	3.4
Gallium Oxide	GaO	4.8
Diamond	C	5.5

FIGURE 6.9 Characteristics comparison of Si, SiC, and GaN.

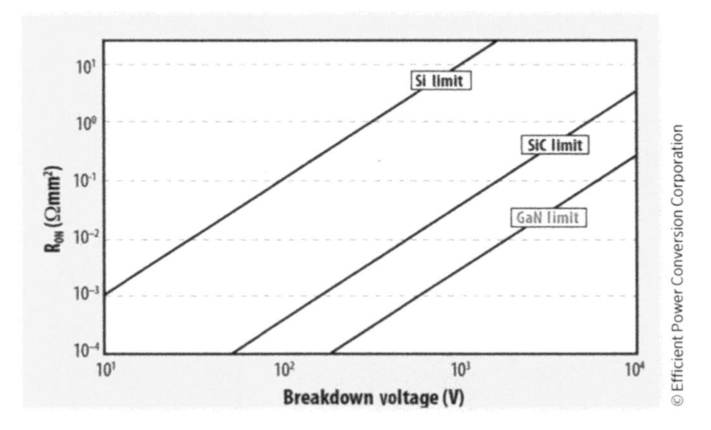

Although WBG semiconductors now cost more than silicon devices, they may eventually be competitive as manufacturing capabilities improve and market applications grow. Several challenges must be addressed to make WBG materials more cost-effective, including:

- Production of larger-diameter WBG wafers.
- Use of novel designs that exploit the properties of WBG materials.
- Use of alternative packaging that allows higher temperature WBG devices.
- Design of systems that integrate WBG devices so they take advantage of their unique capabilities.

GaN and SiC semiconductor materials allow smaller, faster, more reliable devices with higher efficiency than their silicon-based cousins. These capabilities make it possible to reduce weight, volume, and life cycle costs in a wide range of power applications. Figure 6.9 compares the breakdown voltage and on-resistance of Si, SiC, and GaN devices.

Gallium oxide (GaO) is another semiconductor material with a wide bandgap. GaO has poor thermal conductivity, but its bandgap (about 4.8 eV) exceeds that of SiC, GaN, and Si. However, GaO will need more R&D before becoming a major power system participant.

6.10 Silicon Carbide (SiC)

Compared to Si, SiC has:

- On-resistance up to two orders of magnitude lower.
- Reduced power loss in power conversion systems.
- Higher thermal conductivity and higher temperature operation capability.
- Enhanced performance due to material advantages inherent in its physical properties.

SiC excels over Si as a semiconductor material in 600 V and higher rated breakdown voltage devices [21–23]. SiC Schottky diodes at 600 and 1200 V ratings are commercially

FIGURE 6.10 SiC power modules and diodes (with "APT" designations) are included in the power stage of a prototype EV battery charger developed at North Carolina State University. Semikron diodes are high-power silicon devices.

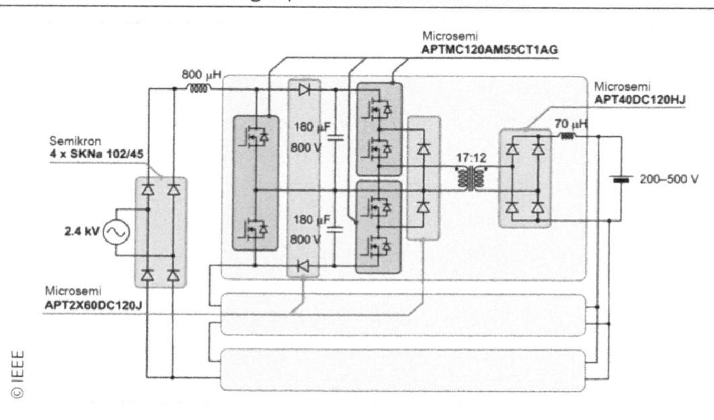

available and accepted as the best solution for efficiency improvement in power converters. Figure 6.10 is an example of an EV battery charger that employs SiC power devices. This charger was developed at North Carolina State University [24].

A design barrier for SiC is low-level parasitics. If there are too many internal and external parasitics, their performance can decrease to that of a silicon device or circuit malfunctions. To break through this barrier, power designers may have to adopt new design and measurement techniques [25, 26].

Conducted EMI can accompany the fast voltage and current-switching transients produced by SiC MOSFETs. Internal and external SiC parasitics are affected by these switching transients and are the primary cause of EMI. System packaging configurations can aid EMI reduction, such as a three-dimensional spatial layout that utilizes multilayer PCB technology and use of SMT components. You can also block EMI with barriers of conductive materials. This shielding is typically applied to enclosures to isolate electrical devices from their surroundings and to cables to isolate wires from the environment through which the cable runs.

SiC MOSFETs are available as a 1200 V, 20 A device that has a 100 mΩ on-resistance at a +15 V gate source voltage. Besides the inherent reduction in on-resistance, SiC also offers a substantially reduced on-resistance variation over operating temperature. From 25°C to 150°C, SiC variations are in the range of 20% versus 200% to 300% for Si. The SiC MOSFET die is capable of operation at junction temperatures greater than 200°C but are limited by its TO-247 plastic package to 150°C. The technology also benefits from inherently low gate charge, which allows use of high switching frequencies, allowing smaller inductors and capacitors [27].

6.11 SiC Automotive Applications

Adoption of SiC-based power solutions is growing across the automotive industry as it accelerates its move from ICEs to EVs. Industry sources estimate that by 2030, 30 million high-voltage electrified light vehicles will be sold representing 27% of all vehicles sold annually.

A recent example is an 800 V SiC inverter being developed by Delphi Technologies that will significantly extend EV range and cut charging time in half compared with today's 400 V inverters. Other advantages will include a smaller battery, lighter, cheaper cables; and greater harvesting of vehicle kinetic energy when braking, which extends vehicle range. These inverters will utilize Cree's Wolfspeed® SiC MOSFETs.

At the heart of the new Delphi Technologies inverter is its patented Viper power switch that combines high levels of integration with unique double-sided cooling. These critical features allow inverters that are 40% lighter and 30% more compact than other inverter designs.

The new SiC Viper power switch will fit into the same inverter package as Delphi's existing silicon IGBT switch version. The IGBT Viper power switch includes a diode and fits in a single, electrically isolated package that is thermally conductive on both sides. Inside each inverter is a power module, which usually consists of six Viper switches that are assembled into one housing. Double-sided cooling reduces power module heat and delivers better reliability in a more compact design.

Viper switch designs reduce the size of the inverter's semiconductor area by up to 50%, saving space and cost. It also permits much higher power outputs, which increases electric vehicle driving range on a single charge.

This Viper power switch eliminates wire bonds-the complex web of wiring that connects the silicon die to the inverter's lead frame. Wire bonds typically are a leading cause of failure in current designs. No wire bonds result in better reliability and durability.

6.12 Gate Drive Layout

For fast switching, the WBG gate drive interconnections must have minimum parasitics, especially inductance, therefore:

- Locate the gate driver as close as possible to the SiC MOSFET.
- Exercise care selecting an appropriate external gate resistor to manage voltage overshoot and ringing.
- Use ferrite beads to minimize ringing while maintaining fast switching speed.
- Use a high value resistor (10 kΩ) between gate and source to prevent excessive floating of the gate during system power-up propagation delays.

6.13 Gallium Nitride (GaN)

The temperature coefficient of GaN FETs on-resistance is positive like the silicon MOSFET, but the magnitude is significantly less. Compared to a silicon semiconductor, GaN's higher electron mobility enables a smaller size device for a given on-resistance and breakdown voltage. This allows GaN devices to be physically smaller and have electrical terminals closer together for a given breakdown voltage requirement [25, 28, 29].

Silicon power MOSFETs have not been able to keep pace with changes in the power electronics applications. Many applications have reached silicon MOSFET's theoretical performance limit, so they must go to another semiconductor material, such as GaN.

The threshold of enhancement mode GaN FETs is lower than that of silicon MOSFETs. This is made possible by the almost flat relationship between threshold and temperature along with the very low gate-to-drain capacitance (C_{GD}). The device starts to conduct significant current at 1.6 V, so care must be taken to ensure a low impedance path from gate-to-source when the device needs to be held off during high-speed switching [30, 31].

FIGURE 6.11 Depletion Mode GaN characteristics.

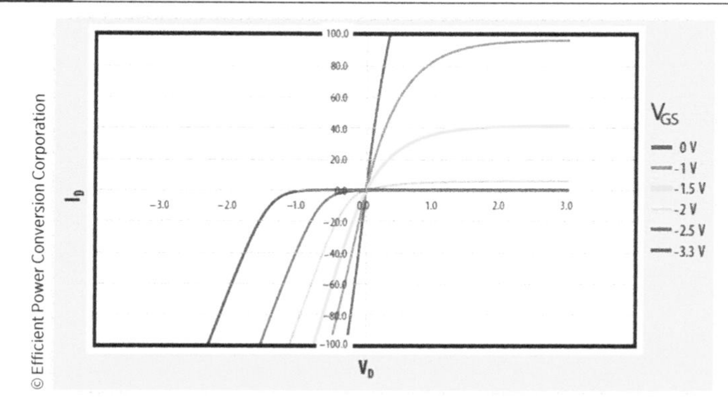

FIGURE 6.12 Enhancement mode GaN characteristics.

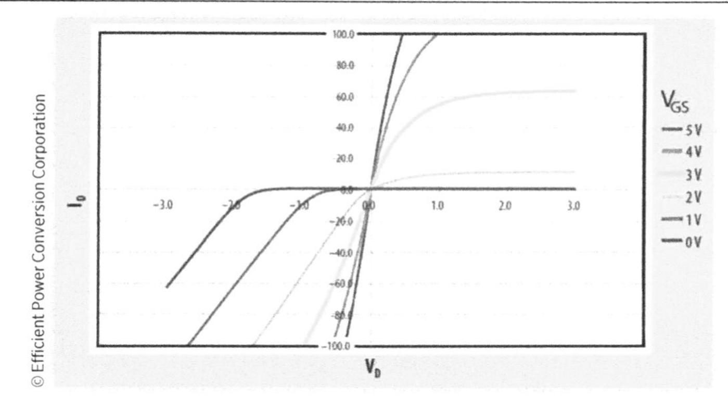

GaN transistors may be depletion or enhancement mode. The depletion mode transistor is normally on and is turned off with a negative voltage relative to the drain and source electrodes (Figure 6.11). In contrast, the enhancement mode transistor is normally off and is turned on by positive voltage applied to the gate (Figure 6.12). Depletion mode transistors are inconvenient because at start-up of a power converter, a negative bias must first be applied to the power devices or a short circuit will result. Enhancement mode devices do not have this problem: with zero bias on the gate, an enhancement mode device is off and will not conduct current.

Initially, GaN-on-silicon transistors were depletion mode types. They operated like a normally on power switch that required a negative voltage to turn them off. However, the ideal mode for designers is an enhancement mode transistor that is normally non-conducting and requires a positive voltage to turn it on. Efficient Power Conversion (EPC) produces an enhancement mode, eGaN, transistor using a proprietary process with a GaN-on-silicon structure. In operation, a positive gate voltage turns the enhancement mode GaN transistor on.

Transphorm employs a cascode configuration with a depletion mode GaN (Figure 6.13). This provides the ruggedness of a silicon gate, coupled with the improved voltage blocking characteristics of a high-voltage GaN. There are no special requirements for the gate driver since the gate is connected to a standard silicon gate rated at ±20 V with threshold around 2 V [32–35].

FIGURE 6.13 Transphorm depletion mode GaN transistor uses a cascade configuration.

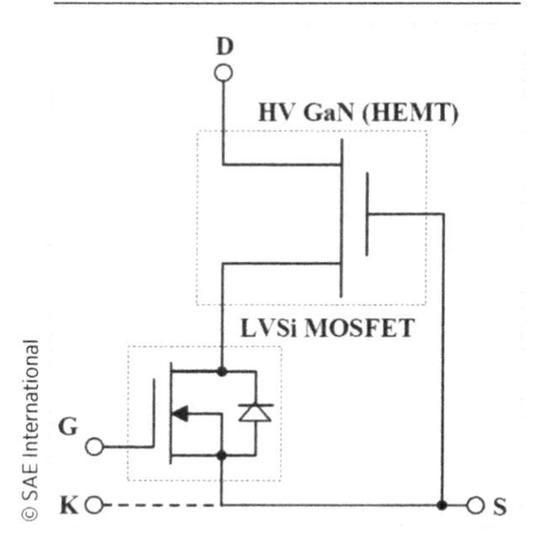

6.14 GaN Drivers

Important parameters for driving a GaN FET are:

- Maximum allowable gate voltage
- Gate threshold voltage
- Body diode voltage drop

The maximum allowable gate source voltage for an eGaN FET of 6 V is low compared with traditional silicon. The gate voltage is also low compared to most power MOSFETs, but does not suffer from as strong a negative temperature coefficient. And, the body diode forward drop can be a volt higher than comparable silicon MOSFETs.

Because the total Miller charge (Q_{GD}) is much lower for an eGaN FET than for a similar on-resistance power MOSFET, it is possible to turn on the device much faster. Too high dv/dt can reduce efficiency by creating shoot-through during the "hard" switching transition. It would therefore be an advantage to adjust the gate drive pull-up resistance to minimize transition time without inducing other unwanted loss mechanisms. This also allows adjustment of the switch node voltage overshoot and ringing for improved EMI.

Virtually all commercially available GaN power semiconductors are now lateral devices that can't handle over about 600 V. New research is focused on producing a vertical GaN structure that is better for medium- and high-power applications. These are devices where the current, instead of flowing through the surface of the semiconductor, flows through the wafer, across the semiconductor.

Vertical devices are better in terms of how much voltage they can manage and how much current they control. Current flows into one surface of a vertical device and out the other. That means that there's simply more space in which to attach input and output wires, which enables higher current loads. Advantages of vertical GaN FETs could include:

- Device dimensions (and cost) roughly independent of breakdown voltage
- Vertical current extraction >100 A possible
- Uniform heat generation and dissipation
- Reduced dynamic on-resistance and current collapse
- New flexibility for normally off device technology
- Ultralow leakage current
- Potentially better reliability as E-field far from surface
- Avalanche breakdown demonstrated
- Drift region could be graded for optimum critical field

Among the current AEC-qualified GaN MOSFET suppliers are:

- Transphorm
- Efficient Power Conversion

Among the announced AEC-qualified SiC suppliers are:

- ON Semiconductors
- Wolfspeed
- Microsemi
- Rohm

References

1. Electronics Tutorial, "MOSFET as a Switch," Electronics Tutorials.
2. Wikipedia, "Power MOSFET."
3. Toshiba, "Power MOSFET Structure and Characteristics," Toshiba Application Note, July 26, 2018, 1-5.
4. Alpha & Omega Semiconductor, "Power MOSFET Basics," Alpha & Omega Semiconductor Application Note, November 10, 2009, 1-10.
5. Davis, S., "Power Semiconductors," powerelectronics.com.
6. STMicroelectronics, "MDmesh M5 Series," STMicroelectronics.
7. STMicroelectronics, "STripFET F7 Series," STMicroelectronics.
8. ROHM, "Power MOSFET Basics: Understanding Superjunction Technology," ROHM.
9. Korec, J. and Bull, C., "History of FET Technology and the Move to NexFET™," Texas Instruments, March 28, 2011.
10. Texas Instruments, "FemtoFET Power MOSFETs," Texas Instruments.
11. Infineon, "DirectFET," Infineon.
12. Qin, J., "Using the Right Power Semiconductors Can Increase Your Power Density," International Rectifier, March 20, 2014.
13. Vishay, "PolarPak," Vishay.
14. Havanur, S., "Designing with DrMOS, Part 1: Concept and Features," Power Electronics Technology 37, no. 12, 26–30 (2012).
15. Wikipedia, "Moisture Sensitivity Level."
16. Liu, Z., Mei, W., Zeng, X., Yang, C. et al., "Remaining Useful Life Estimation of Insulated Gate Bipolar Transistors Based on Novel Volterra k-Nearest Neighbor," MDPI, November 3, 2017.
17. Laud, S., "IGBT FAQs," powerelectronics.com, May 6, 2014.
18. Toshiba, "IGBTs," Toshiba App Note-40, September 1, 2018.
19. Murata, "Powering IGBT Gate Drives with DC-DC Converters," Murata Application Note, May 2014.
20. Wikipedia, "Wide Bandgap Semiconductors."
21. Davis, S., "SiC Transistor Basics," Power Electronics Technology, October 9, 2013.
22. Keim, R., "Exploring Pros and Cons of Silicon Carbide (SiC) FETs," All About Circuits, March 28, 2017.
23. Tech Web, "What Are SiC MOSFETs? Comparison of Power Transistor Structures and Feature," Tech Web, August 10, 2017.
24. Davis, S., "SiC Power Devices Lead to More Efficient, Smaller EV Battery Charger," powerelectronics.com, November 16, 2018.
25. Lapedus, M., "What Happened to GaN and SiC? Semiconductor Engineering," February 2, 2015.
26. Rabkowski, J., Pefitisis, D., and Nee, H., "Silicon Carbide Transistors: A New Era in Power Electronics Is Initiated. IEEE Industrial Electronics Magazine 6, no. 2 (2012): 17-26.
27. Singh, R., "Silicon Carbide Switches in Emerging Applications," GeneSiC Semiconductors Inc., September 17, 2015.

28. Lidow, A., Strydom, J., de Rooij, M., and Ma, Y., GaN Transistors for Efficient Power Conversion (El Segundo, CA: Power Conversion Publications, 2012).
29. Brohlin, P., Ramadass, Y., and Kaya, C., "Direct-Drive Configuration for GaN Devices," Texas Instruments, November 20, 2018, 1-7.
30. Dusmez, S., Xie, Y., Beheshti, M., and Brohlin, P., "Thermal Considerations for Designing a GaN Power Stage," Texas Instruments, October 23, 2018.
31. Efficient Power Conversion, "What Is GaN?," Efficient Power Conversion.
32. Wikipedia, "Gallium Nitride."
33. Electrical Engineering, "Why Don't We Use GaN Transistors Everywhere?," Electrical Engineering Data Sheet, June 5, 2016.
34. Beach, R., "Master the Fundamentals of Your Gallium Nitride Power Transistor," Electronic Design, April 29, 2010.
35. Davis, S., "GaN Basics: FAQs," Powerelectronics.com, October 2, 2013.

CHAPTER 7

EV Lighting

Power management of lighting systems is evolving as more light emitting diodes (LEDs) are used for automotive applications. This evolution has been aided by the development of efficient LEDs that emit a white light instead of their original red light. Power management controls LED light output, allowing them to produce more light for their power input than a conventional incandescent lamp. Initially used for internal combustion engine automobiles, EVs are following their lead with LED lighting that minimizes lighting power requirements and allows longer driving range for a single charge [1].

7.1 LED Basics

LED development was based on the finding that light can be produced by causing an electron in a material to jump from its valence band to a conduction band (bandgap). This energy will eventually fall back into the hole it left in the valence band releasing energy back again in the form of a photon (light).

The two main types of LEDs presently used for lighting systems are aluminum gallium indium phosphide (AlGaInP or AlInGaP) alloys for red, orange, and yellow LEDs and indium gallium nitride (InGaN) alloys for green, blue, and white LEDs. The output from an LED can range from red (at a wavelength of approximately 700 nm) to blue-violet (about 400 nm) [2].

Using an ultraviolet or violet LED packaged inside a phosphor-coated enclosure, you can produce a full-conversion white LED. Full-conversion LEDs are inefficient, so manufacturers went to partial conversion LEDs that illuminate the phosphor with blue light, converting a portion of that blue light into white light. Typical interior applications for these LEDs are:

- Cluster or instrument backlighting
- Dome or map reading lights
- Courtesy lights

Converting to LED lighting provides advantages of:

- Higher reliability and long service life compared with incandescent lamps.
- Fast turn-on time for stop lights because light output rises to full intensity approximately 250 ms faster than incandescent bulbs.
- Small size enables flexibility in package design and overall appearance.
- Better vibration resistance than incandescent lamps.

LEDs in automotive applications must comply with guidelines for temperature and humidity range, ability to withstand adverse environments, electromagnetic interference (EMI), and voltage protection circuitry. Also, automotive components must pass reliability requirements dictated by qualification testing. For automotive use LEDs must qualify for AEC-Q102 and LED driver ICs must comply with AEC-Q100. Because of their advantages, LED lighting applications in cars has been growing.

Light output levels from packaged LED devices has roughly doubled every 18 months. With the increased efficiency and lower costs, more LEDs and driver ICs are being used in vehicles. The technology of power management functions in LED driver ICs has also advanced, allowing much more efficient control of LED lighting.

Exterior applications for white LEDs include:

- Headlights
- Taillights
- Turn signals
- Brake lights
- Parking lights
- Side marker lights
- Fog lamps
- Daytime running lights

7.2 Headlights

LED headlights are a major advantage for EVs. They could offer an 85% reduction in energy consumption compared with incandescent bulbs, which translates into increased driving range. Also, new high-brightness (HB) LEDs (HBLEDs) are being used more extensively as headlights [3].

Limiting factors of LED headlights include high system expense and temperature limitations. Because of their brightness requirements, headlight LEDs need higher power than interior LEDs. A headlight LED's performance depends on its temperature, because it will produce more light at a low temperature than at a high temperature.

Maintaining a constant light output regardless of temperature is necessary for an LED headlight.

Heat built up within LEDs can be an issue in headlights, requiring thermal management for the LEDs and their associated driver ICs [4]. Additionally, LEDs generate considerable heat per unit of light output and are damaged by high temperatures. Prolonged operation above their maximum junction temperature will permanently degrade LEDs and ultimately shorten their life. Maintaining LEDs at a low enough junction temperature at a headlight's high power levels requires effective thermal management, such as heat sinks or cooling fans, which add to automotive system cost.

Sometimes it is not convenient to install an adequate heat sink in the headlight housing, so the temperature must be controlled electronically using a sensor to detect operating temperature and a controller to reduce the LED brightness until the temperature returns to a safe value. Pulse-width modulation (PWM) could be used to control LED brightness and operating temperature.

There are also thermal management issues with LED headlights in cold, ambient temperatures. Excessively low temperatures decease LED light output beyond the regulated maximum. LED lifetime may not be reduced at low temperatures, just light output is reduced. Lamp temperature can be monitored, and the lamp could be heated whenever it is sensed that the temperatures is too low.

An LED headlight must be insensitive to vibration and shocks and needs to be operated within certain temperature limits. Small fans specially adapted to the needs of headlight are one possible solution.

Most headlights employ multiple LEDs. If even one LED fails, the entire headlight must be replaced, so all components must be durable and perform reliably. Even in an energy-saving, LED headlight housing with several LED chips, the amount of heat to be dissipated adds up to a few watts. If the chip gets too hot, the service life and light output both suffer. Small fans that are powerful, yet smooth-running and durable, help to direct the cooling airflow with pinpoint accuracy to protect the electronics. Ruggedly designed fans that can stand up to vibrations, shocks, and heat and cold (–40°C to +125°C) are required in a closed headlight housing.

7.3 High-Brightness LEDs

HBLEDs are a new generation of LEDs that are bright enough for exterior illumination applications, such as the headlight. These HBLEDs offer much higher levels of luminosity than standard LEDs. One of the chief reasons for using HBLEDs is their improved efficiency over other types of lamps [5].

HBLEDs produce over 50 lumens of light [6]. They are energy efficient, eco-friendly, low power, and longer lasting than the average incandescent bulbs. HBLEDs have multiple advantages over most LEDs:

- Brighter
- Longer life
- Relatively low cost
- Restriction of hazardous substances (RoHS) manufacturing
- Precision designed optics, multiple distributions, lumen outputs, and color temperatures make the HBLED ideal for headlight applications

7.4 LED Drivers

The requirements for headlight LED drivers are much more complex than driver ICs for interior lighting. It would be best to use a switch-mode voltage regulator with external MOSFETs for headlight lighting, rather than a power supply with integrated MOSFETs, to minimize heat dissipation. Besides supplying extra power, the headlight LED driver IC should be able to handle dimming requirements for headlights. Normally, the exterior lighting of a car has to be dimmed by varying the brightness of each individual LED, such as for brake lights/taillights and high-beam/low-beam. These are called bilevel settings, and the power requirements put extra strain on the voltage converter/driver. Sometimes, two separate drivers are needed for the two settings and, at other times, one LED driver can address both situations.

Extensive protection and fault detection functions are required to prevent LED driver IC failure. To ensure high reliability in automotive applications, driver ICs should be protected against overvoltage, undervoltage, reverse polarity, overcurrent, short circuits, and overtemperature.

LEDs require a constant current to produce consistent lighting so it establishes the basic operating requirements for the associated driver. Current fluctuations occurring with voltage supply variations in vehicles should be avoided. Linear regulators provide a simple control and do not require EMI filters. However, their power dissipation can become excessive for high power applications [7–9].

The ability to adjust light intensity in interior lighting systems is a normal automotive requirement. For exterior lighting, the same LED may provide different levels of brightness. For example, the stop and position lights, or low-beam and high-beam headlights, are defined at two brightness levels. Also, brake lights/taillights, low-beam/daytime running lights, and high-beam/low-beam headlights are so-called bilevel lightings. In some cases, lighting design may be able to address both situations with the same LED by using an appropriate LED driver.

7.5 Dimming LEDs

LEDs can be dimmed in two ways: analog and PWM dimming. Analog dimming changes LED light output by simply adjusting the DC current in the string, whereas PWM dimming provides pulses of full-amplitude current to the LEDs. The driver varies the duty cycle of the pulses to control the apparent brightness. A key factor in dimming is the frequency that determines the eye's sensitivity to flicker. PWM dimming relies on the capability of the human eye to integrate the average amount of light in the pulses. Provided the pulse rate is high enough (typically about 200 Hz), the eye does not perceive the pulsing but only the overall average. PWM dimming is generally more complex to implement than analog dimming, but it maintains high efficiency and ensures the LED light output does not vary in color [10].

Immunity to EMI is an important consideration for all automotive electrical components. Switching the current on and off through the LED at high PWM dimming pulse rates can radiate EMI. To control EMI during PWM dimming, the rise times and fall times of the pulses are increased.

The basic linear regulator would provide a simple control while avoiding the need for EMI filters. However, a linear regulator exhibits higher power dissipation than a switch-mode regulator when used to supply LED headlight power. A switch-mode buck regulator can be used to avoid the higher power dissipation. When higher voltage outputs are needed, a buck–boost topology can be used when the driver controls several LEDs in series. A buck–boost voltage regulator can handle both the varying supply voltage and the output voltage variation [9].

7.6 AEC-Q102 Rev—March 15, 2017 [11]

AEC-Q102 defines the minimum stress test-driven qualification requirements and references test conditions for qualification of discrete LEDs in all exterior and interior automotive application. It combines state-of-the-art qualification testing documented in various norms (e.g., JEDEC, IEC, MILSTD) and manufacturer qualification standards.

For the qualification of parts using optoelectronic functions together with other components (e.g., multichip modules with sensors and integrated signal processing, solid-state relays, LEDs mounted on boards with additional mechanical connectors, etc.), it is mandatory to combine tests defined in this specification with further tests described in other adequate (AEC) norms.

Successful completion and documentation of the test results from requirements outlined in this document allows the supplier to claim that the part is "AEC-Q102 qualified." The supplier, in agreement with the user, can perform qualification at sample sizes and conditions less stringent than what this document requires. However, that part cannot be considered "AEC-Q102 qualified" until the unfulfilled requirements have been successfully completed.

For electrostatic discharge (ESD), it is highly recommended that the passing voltage be specified in the supplier datasheet with a footnote on any pin exceptions. This will allow suppliers to state, for example, "AEC-Q102 qualified to ESD H1B," implying that supplier passes all AEC tests except the ESD level. Note that there are no "certifications" for AEC-Q102 qualification and no certification board run by AEC to qualify parts.

The minimum temperature range for discrete optoelectronic semiconductors per this specification shall be −40°C up to the maximum operating temperature defined in the part specification.

7.7 Test Requirements and Failure Criteria for LEDs

For a pre- and post-stress electrical tests, the following electrical and optical parameters shall be used (as a minimum), measured at room temperature:

- Luminous flux or intensity or radiant power (whatever is appropriate)
- Color coordinates Cx and Cy or dominant wavelength (for direct colors)
- Forward voltage V_f ±10%
- Forward voltage V_{min} ±10%
- Forward voltage V_f (light/no light)

LED lifetime depends on the specific application; interior lighting has different requirements than exterior rear and exterior front lighting. Also, trucks may have different requirements than the majority of personal cars.

Reliability validation for LEDs requires additional measurements. Determining failure modes involves overstress tests. The following tests derived from SAE/USCAR-33 are recommended:

- High temperature operating life (HTOL)
- High humidity and temperature operating life
- Power temperature cycle
- Temperature shock

7.8 Volkswagen Sees the Light

"We are working hard to improve safety in night driving situations," said Mathias Thamm, head of the Technologies and Innovations subdepartment of the Volkswagen brand. This is necessary because, while around 30% of all accidents involving personal injury occur at night, the severity of the accident is about twice as high as during the day [12].

Early generations of Volkswagen's Golf had halogen headlights that became brighter and brighter. These were followed by the first xenon headlight, which revolutionized light. Then came LED tail lamps, LED daytime running lights, and LED headlights. Now, Volkswagen has the IQ.Light, which it considers one of its best lighting systems.

Volkswagen partnered with German-based auto parts producer Hella to develop the IQ.Light LED headlight. These adaptive matrix headlights (Figure 7.1) employ a targeted control of up to 128 LEDs to provide precise and optimal light distribution and luminous intensity. Volkswagen contends that, compared to conventional lighting systems, the LED Matrix Headlight will enhance safety and comfort because they will allow drivers to recognize road details and possible obstacles at night earlier, giving them more time to respond.

The headlight has individually controlled LEDs that blend into a matrix of light areas in both high and low-beam modules. The low-beam employs a matrix of 48 LEDs positioned on a shared circuit board. The high-beam circuit board uses 27 LEDs. Combined, these 75 LEDs form the adaptive matrix light.

FIGURE 7.1 Volkswagen's IQ.Light LED adaptive matrix headlight.

© Volkswagen AG

Another 53 additional LEDs ensure that lighting functions can be shown, including the illumination in front of the vehicle, as well as the daytime running lights, the cornering lights, the position lights, and the direction indicators. In total, the front lamps of the new Volkswagen Touareg will use 256 LEDs.

Assisting the headlight is a front digital camera that analyzes road conditions, spots oncoming vehicles, and so on. That data combined with GPS data, along with speed and steering angles, will go into selecting the LEDs in the matrix to provide the ideal headlight illumination for the road and the surrounding area up ahead in less than a second.

The camera-based system reacts to signs by temporarily dimming the LEDs to avoid causing glare for drivers themselves. The camera can't detect the wet road, so the driver has to activate the bad-weather light function manually.

The new technology creates the following possibilities, for example:

- **Sign glare control**: A camera can detect a traffic sign or a warning sign, and the IQ.Light headlight dims in milliseconds so these signs no longer reflect so intensively, impairing the driver's vision. This is particularly important for large-format directional signs at freeway interchanges.
- **Direction indicator**: Projects arrows on the road to indicate change of direction to an oncoming car (Figure 7.2).
- **Optical lean assistant**: The light projects the lane width onto the road, for example, on roads narrowed by roadworks. This enhances safety, particularly at night.
- **Dynamic light assistant**: This feature, which already comes in many vehicle series, allows the driver to drive in high-beam mode at all times. If another vehicle or other road user approaches, the headlights dim automatically in milliseconds.
- **Infrared**: The integrated night camera onboard detects a person or animal at the edge of the road further along and immediately turns on individual headlight of the IQ.LED matrix headlight that illuminate the person's or animal's exact position on the road. The driver sees what is ahead very clearly. People or animals at the edge of the road are detected by infrared technology and briefly illuminated.
- **LED matrix lamps**: In the future, they could, among other things, display warning notices in the tail lamps to defuse dangerous situations, such as the end of a tailback via car-to-car communication.
- **Optical Park Assist**: New assistance functions such as "Optical Park Assist," which works with micro lenses, will make maneuvering in narrow parking spaces and parking garages easier and safer.

FIGURE 7.2 Arrows projected on the roadway indicate change of direction in Volkswagen automobile.

Klaus Bischoff, Volkswagen's designer-in-chief, noted:

> *The lighting of the future will become a means of communication. It will interact with the driver and with other road users—whether in a car, on a motorcycle or bicycle or as a pedestrian on the road—measurably improving safety. At the same time, we will integrate the lighting functions into the design of the vehicles more progressively than ever before.*
>
> *What now often works through eye contact, for example, at pedestrian crossings, is not possible with autonomous driving. If the driver is no longer on board or is otherwise busy in the car, the car must communicate directly with the pedestrian. This can take place by means of light signals at the front of the vehicle that signal to the pedestrian that the car has recognized him and he can walk across the road.*

In order to be perfectly prepared for the challenges of the future, Volkswagen opened its own lighting competence center at the Wolfsburg, Germany, plant in 2014. There, a 100-m long, 15-m wide, 5-m high light tunnel has since been in operation in the middle of the R&D department. In this tunnel, the lighting systems of today and tomorrow are tested on a real road simulation. The tests can be repeated and reproduced with total precision. Systems can be compared and evaluated in a light tunnel better than ever before.

The competence center is also an ideal place to investigate the light perception of drivers and passers-by. Also, interior systems such as ambient lighting, heads-up displays, and infotainment systems can be tested here under reproducible conditions. The light tunnel also shortened the development time for new headlight, tail lamp, and interior lighting systems, as the number of time-consuming night drives could be reduced. This allows advances in lighting development to be implemented even more quickly into series production technologies, such as the new IQ.Light, providing additional safety to the benefit of all road users.

References

1. Benchoff, B., "History of White LEDs," Hackaday, October 29, 2018.
2. Leung, M., "White LED and Remote Phosphor Comparison," Cree, 2014, 1-10.
3. Ling, O.S., "Thermal Modeling of High Power LEDs," Avago, 2010, 1-5.
4. ebmpapst, "Thermal Management for LED Headlights," ebmpapst, September 16, 2013.
5. Nexperia, "Focus MOSFET Application – Automotive Lighting," Nexperia, 1-2.
6. Hughes, L., "Beyond HID Headlights: The Future of Automotive Lighting," Arrow, April 26, 2017.
7. Maxim, "Selecting HB LED Drivers for Automotive Lighting," Maxim, March 4, 2009, 1-5.
8. Li, S., "How a Multiphase Driver for Headlamp Light Control Works," NXP, November 15, 2016, 1-2.
9. Infineon, "Automotive LED Lighting," Infineon.
10. Lange, F., "LED Drivers for Automotive Applications," EE Times, April 10, 2012.
11. AEC, "AEC - Q102," AEC, March 15, 2017.
12. Davis, S., "The Future Looks Bright for Innovations in Automotive Lighting," powerelectronics.com, December 4, 2018, 1-4.

CHAPTER 8

EV Traction Motors

Power management controls the power applied to the traction motor so it can propel the EV. Most EVs employ an AC-powered motor, so power management converts the EV's primary DC power source to the proper AC voltage and current for the traction motor. Power applied to the traction motor produces a magnetic field that causes the motor's shaft to rotate.

8.1 Electric Motors vs. Gasoline Engines

"Electric motors are inherently superior to gasoline engines," said David Grace, a former Tesla drivetrain developer. Grace said that there are several characteristics that give electric motors advantages over gasoline engines [1]:

- Electric motors generate motion, not heat. A fossil fuel engine produces motion, basically, with tiny controlled explosions. Those explosions push interlocking pieces of metal that connect to a driveshaft. All that metal rubbing together generates a lot of heat, even when the parts are swimming in oil. That's energy not being used to push a vehicle forward.
- There is zero contact between the electric motor and a vehicle's driveshaft, just an air gap. With the driveshaft pushed magnetically instead of mechanically, even a running electric motor is barely warm to the touch, eliminating a major energy waste.
- Wasted heat from a gas engine becomes free cabin heat in the winter, while electric vehicles must produce extra heat as needed. Cold also has a negative impact on

batteries, meaning that electric systems take hits to range, performance, and charging speed in the cold.

- At lower speeds, electric motors deliver more torque than gas engines. "Torque is what you need to get a car going," says Grace, and, with more of it, electrics out-accelerate comparable gas engines. Gas engines, however, do perform better at very high speeds than electric motors.
- Electric motors are simpler. Electrics' better torque has a secondary advantage. With less torque at low speeds, gas engines need help from a transmission to get moving. Electric engines by and large don't need transmissions.
- Electric motors are easier to service. Electric drivetrains have far fewer subsystems; no transmission, oil tank, or catalytic converter. That means there's less to break down. The electric motor is much smaller and more streamlined than its gas counterpart.
- Every electric motor is also an electric generator. The rotor is always spinning in the same direction, but the electrical field reverses. That sends electrons streaming back into the battery while helping slow the vehicle with regenerative braking.
- Electric motors are smarter and provide more and better opportunities to monitor and adjust them.
- Electric motor control systems are much more accurate and probably more transparent as to what's going on. As software updates become an increasingly regular part of car ownership, electrics will be that much more flexible.
- Better metrics also let manufacturers detect faults in vehicles before becoming a big problem. The support team can arrange a fix before the customer notices anything is wrong.

The conversion of vehicles from gasoline engines to electric motors requires new technology and background information to support this conversion, including:

- Higher power density, safer batteries
- Lower temperature, more efficient fuel cells
- Public and private battery recharging facilities
- Effective and safe "recharging" of fuel cells with hydrogen
- Teaching operating personnel and repairmen how to handle high voltages for electric motors, particularly after an accident

8.2 Traction Motors

EV traction motors have some special requirements that set them apart from conventional electric motors whose basic design has not changed since it was invented almost two centuries ago [2]. For example, the electric motor design determines how much electrical energy from the battery is transformed into physical energy used to turn the wheels of the vehicle in an EV. Therefore, maximizing efficiency is one of the most important challenges in motor design. Minimizing electrical magnetic losses is critical, since any loss leads directly to shortened battery run time. The loss also generates heat, which must be removed from critical components, possibly resulting in further energy losses via fans and cooling flow modifications. Copper loss and lamination core losses must be determined at a wide range of operating conditions, such as speed and current [3].

The traction motor must also recharge the battery via regeneration braking, so motor efficiency is critical to this role as well. Any loss means that energy captured from the decelerating vehicle is not fully absorbed into the battery.

Reliability is also important for automobile owners. It is a critical element in controlling warranty costs and brand image. Traction motors must operate consistently under numerous extremes: hot and cold temperatures, severe vibrations, hard-duty cycles, and rough road conditions. Plus, the traction motor is exposed to high temperatures produced under the hood. These and many other variables must be addressed during the motor design process.

There are also tradeoffs in thermal design. The power delivered to the wheels as well as the power needed to recharge the battery travels through the power electronics. Even the slightest power loss could create large amount of heat. The heat must be carefully managed and dissipated under a multitude of operating conditions, ranging from desert to subzero winter conditions, to avoid damage to the power electronics and nearby components. Therefore, motor designers must accurately calculate electric losses and identify/design heat dissipation paths to ensure effective cooling.

One type of EV motor is the AC induction type. These motors use electric power to generate electromagnetic fields inside the motor that rotates the rotor. Another EV motor is the permanent magnet type that doesn't need that additional current because its magnets are always providing the necessary magnetic fields. The permanent magnet synchronous motor (PMSM) is more efficient, making it better for smaller and lighter cars, but not ideal for high-performance cars, because an AC induction motor can produce greater power [4, 5].

8.3 AC Induction Motors

Most electric vehicles use either an AC induction motor or PMSM. The induction motor induces magnetism that is leveraged to output rotary motion. The stationary outer stator is connected to an external electrical power source. This is fed to the rotor's poles in a rotating progression that causes revolutions of the magnetic field within the motor. Conducting bars in the rotor interact with the stator's magnetic fields; current is induced in those bars, which, in turn, generate magnetic fields that are attracted to those of the stator. Because an AC induction motor increases the flux enclosed by its stationary coils, it is a transformer with a rotating secondary (rotor). The rotor current's effect on the air gap flux causes torque [6].

8.4 Permanent Magnet Synchronous Motors

The PMSM has a permanent magnet rotor and windings on the stator. The stator structure has windings constructed to produce a sinusoidal flux density in the airgap of the machine that resembles an induction motor. PMSM's power density is higher than induction motors with the same ratings since there is no stator power dedicated to magnetic field production. These motors are designed to be more powerful while also having a lower mass and lower moment of inertia.

Venkat Viswanathan, a mechanical engineering professor at Carnegie Mellon University, studies EV performance. He says "The motor efficiency map, that is, its efficiency as a function of torque and speed, determines the energy consumption for consumer vehicles, and the peak power characteristics are an important factor for high-performance demands. In addition, the heating of the motors in-use—at high speeds—is another area with room for innovation and development" [7].

An important advantage for PMSM over the induction motor is lower losses and higher torque density. Also, the PMSM is:

- Cleaner, faster, more efficient
- Less noisy, more reliable
- More compact, efficient, and lighter than induction motor
- Coupled with field-oriented control (FOC) produces optimal torque
- Smooth low- and high-speed performance
- Low audible noise and EMI

The motor drive system is the main power source of EVs, so the efficiency of the drive system will affect mileage endurance. Compared with induction motors, PMSMs, with their advantage of high power density and high efficiency, can decrease the energy consumption and improve the operational mileage of EVs. Therefore, the PMSM direct-drive system has been widely applied in EVs for transportation. There are several PMSM control strategies, such as maximum torque per ampere (MTPA) control, maximum speed per voltage (MSPV) control, and unity power factor.

The PMSM can offer other advantages, including:

- High power-to-weight ratio
- High efficiency
- Rugged construction
- Low cogging torque
- Additional reluctance torque

There are two different high-performance control strategies for the PMSM, namely, the FOC and the direct torque control (DTC). Both are widely used in industry applications. The FOC is implemented in the rotor flux reference frame and needs the continuous rotor position information to implement the coordinate transformation. The DTC is implemented in the stationary reference frame and doesn't need the continuous rotor position information, except for the initial rotor position. But it suffers from high current and torque ripples and variable switching frequency.

8.5 Traction Motor Efficiency

An EV traction motor experiences both mechanical and electrical losses. These losses affect motor efficiency: power out vs. power in [8].

Mechanical losses in AC motors are mainly due to bearing friction and any wind resistance (windage) that opposes the spinning rotor. Windage is one of the largest contributors to total losses. Motors with smooth rotors, such as the permanent magnet (PM) and AC induction types favored in EVs, will suffer less windage losses than comparably sized motors with windings in their rotors. Purely frictional losses are a linear function of rpm and are usually a small fraction of the total losses in the typical (PMSM) or AC induction traction motor.

Electrical losses are "copper" or "iron" in origin. Copper losses include any power consumed by generating the field including the rotor in the AC induction motor, as well as the more obvious resistive loss and the less obvious AC losses (from skin and proximity effects). Skin and proximity effect can be thought of as resistive losses that increase with frequency. Skin effect is the tendency for current to become increasingly constrained to the outer perimeter of a conductor as frequency goes up.

Resistive loss is a copper loss referred to as I^2R loss. It tends to dominate in traction motors because they are usually operated at high currents and low rpms.

The major contributor to iron losses is from hysteresis, which is, basically, the resistance to a change in the direction of magnetization or flux density. Hysteresis loss is a measure of how soft a magnetic material is, and it is most strongly dependent on flux density.

There are various "stray" losses, including magnetic leakage, which is any flux that doesn't link the rotor and stator together, so it doesn't do any useful work. Also, there is unlinked flux that subtracts from the effective AC voltage exciting the armature. There is common-mode, capacitively coupled current that results from the drive rapidly switching its phase outputs between 0 V and bus voltage (typically around 350 V), which causes current to flow across any parasitic capacitances along the way. While the actual power loss from these currents is minimal, they can still erode bearings and damage the insulation on the phase windings. Plus, it may cause the vehicle to fail its EMI test.

8.6 DC-AC Inverter

Figure 8.1 shows a traction inverter that converts the DC from the EV's battery to AC for the three-phase AC induction motor or PMSM that drives the vehicle's propulsion system [9]. It also plays a significant role in capturing energy from regenerative braking and feeding it back to the battery. The PMSM can also be powered by an inverter. However, the PMSM motor requires additional motor controls that are different from those of the AC induction motor.

Common techniques used to control the inverter are pulse-width modulation (PWM) and pulse-frequency modulation (PFM). PWM is the most commonly used technique. PWM varies the width of pulses required for the switching of transistors in an inverter to generate an output waveform composed of many narrow pulses in each cycle. The control unit makes the average voltage of the modulated output pulses sinusoidal.

FIGURE 8.1 Traction inverter converts the DC from the EV's battery to AC for the traction motor.

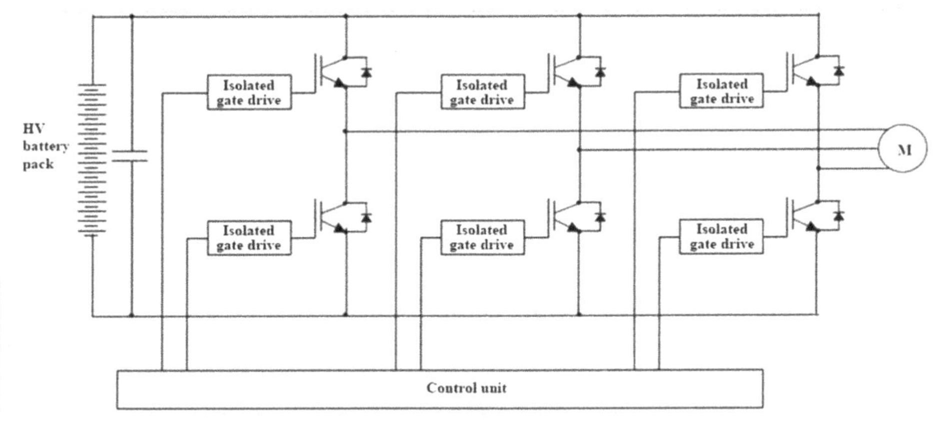

Inverter outputs may provide pure sine wave inverters or modified/quasi sine wave (MSW) inverters. True inverters transform DC into a smoothly varying alternating current very similar to the kind of genuine sine wave. Modified sine wave inverters produce a kind of "rounded-off" square wave, a much rougher approximation to a sine wave that consumes more power than a pure sine wave so there is some risk of overheating with MSW inverters.

There are two different traction inverter design philosophies. The first is separate packaged "box-type" inverters commonly used by many automotive manufacturers and system suppliers. These inverters using power modules benefit from ease of assembly while still having a modular design approach.

The second design philosophy applies a more integrated inverter design where small and fully encapsulated power modules are placed within the mechanical compartment of the drivetrain. Here, the electric machines can be designed as a fully integrated part of the transmission by integrating the power inverter into the same housing.

The demand for higher levels of integration of power modules into inverters is growing and with good reason. The concept of an electrical machine including all the power electronics needed for a frequency inverter, matching requirements for cooling, vibration, and robustness is very attractive when it comes to the sensitive discussion of volume or weight restrictions and overall system cost.

Adequate cooling and ventilation are paramount to keeping the components operational. For this reason, most inverter/converter installations have dedicated cooling systems.

8.7 Future Motors

Future EV motors will have to continue to grow in performance and efficiency. Some innovators will seek out magnets made using more low-cost and non-rare-earth elements, as Honda did in a development project with Daido Steel. Their neodymium magnet contains no heavy rare earth materials but is still powerful enough for vehicle use. Motor speeds will also have to improve; right now, they range from roughly 12,000 to 18,000 rpm, but researchers are developing motors that can reach 30,000 rpm with the benefit that a smaller, lighter motor can do the work of a larger one that spins more slowly.

Improved thermal management will also improve efficiency. In addition, completely new motor designs, such as ultralightweight, in-wheel hub motors, can improve range and performance.

A new axial flux (AF) motor from Magnax may be the key to the high power density motor [10]. The yokeless AF motors have a shorter flux path with lower iron resistance and a small airgap. PMs are further away from the axis compared with traditional motors, resulting in greater leverage around the central axis and thus more torque. Furthermore, the AF design wastes very little copper on overhanging loops on the windings. The motors have zero overhang; that is, 100% of the windings are active. This is not only good for cost but further reduces the weight and copper losses. But the most important element is the fact that, unlike radial flux motors or older AF concepts, the Magnax motor has no stator yoke. This saves an enormous amount of weight and significantly increases the power density. This also means less iron, meaning lower iron losses, which greatly improves the efficiency range of the vehicle. Less iron means less iron losses, which is very important for the no load losses problem (free spinning losses) that we see in multi-motor applications such as the Jaguar iPace. Using yokeless AF motors can easily lead to 10%–20% more efficiency, which also results in lower battery cost.

According to Magnax, a 100-kW AF permanent magnet (AFPM) machine offers the following benefits:

- Increased efficiency (+96% efficiency vs. typical industry values of 84%–92% at lower rpm).
- Reduced axial length (5×–8× shorter than traditional motors/generators).
- Reduced mass (3×–4× lighter than traditional radial flux machines).
- Reduced resources requirement (½–⅓ of materials required vs. other machines, which also results in lower costs).

AFPM machines perform well at a very broad range of rotational speeds, which makes them suitable for high-speed-low-torque and low-speed-high-torque applications.

AF machines are more compact because their axial length is much shorter than radial flux machines, which is often crucial for built-in applications like eAxle and P2 hybrid applications. The slim and lightweight structure results in a machine with a higher power and torque density. Plus, according to the company, AFPM machines can provide the highest energy efficiency of all electrical machines without using exotic materials.

AF designs were confronted with design and production challenges:

- *Mechanical*: The high magnetic forces acting between the rotor and the stator produce an engineering and material challenge in maintaining a high-tolerance uniform air gap between these two components.
- *Thermal*: Windings in an AF machine are located deep within the stator and between the two rotor discs, which presents a greater challenge in terms of cooling than for a traditional radial flux design. The fact that there is no yoke, Magnax had to find other ways to cool the coils.
- *Manufacturing*: AF machines have thus far been very difficult to manufacture because the stator iron's design has continued to be based on that of an RF machine, using a stator yoke to close the flux loop. Magnetic forces between the rotor and stator discs tend to make it very difficult to keep the air gap between them uniform. If they start to wobble or bend, the discs can start rubbing against one another, leading to bearing damage at best, and rapid, spectacular unscheduled disassembly at worst.

Magnax's solution was first to remove the iron yoke from the stator but keep the iron teeth. Then, there are two rotor discs with the stator between these discs. A small air gap exists between rotors and stator that contains the windings while the rotors contain the magnets. The two rotor discs exert an equal (but opposite) attraction force on the rotor.

The discs, however, are directly connected to each other via the shaft ring, so the forces cancel each other out. The internal bearing doesn't carry these forces; it only needs to keep the stator in the middle between the two rotor discs. Theoretically, when the stator is exactly in the middle, it's in equilibrium and no forces act upon the bearing, although there's always a small force that acts on the internal bearing.

Magnax's AF concept is used for electric motor applications where weight and size must be kept to an absolute minimum while delivering high amount of power and torque. Because torque directly relates to diameter, the diameter is an important specification for machine size. Magnax yokeless AF motors can be oil, water, or air-cooled and can easily reach power densities up to 15 kW/kg with fluid cooling.

The peak efficiency of these machines can reach 97% and remains very high at partial loads and lower rpm ranges where the car usually drives. Figure 8.2 shows a Magnax motor installed in an automobile.

FIGURE 8.2 Magnax AXF275 Axial Flux motor that delivers 300-kW with a weight of only 24 kg integrated into a car chassis.

AF motors can be combined with gearboxes in different powertrain configurations, chassis-mounted or in-wheel. Some e-mobility applications require direct-drive motor concepts. A gearless design significantly reduces complexity and maintenance requirements. Just like the in-runner models, these out-runner motors deliver the nominal torque at 0 rpm and have a very compact design, so they're well-suited for direct-drive (in-wheel) configurations. For such applications, they ensure that the efficiency mapping is optimized for lower rpm ranges (usually a wheel speed of 500–2,000 rpm).

8.8 In-Wheel Motors

In-wheel motor systems modify the hub of each EV wheel by adding a complete drivetrain that supplies torque to its associated tire (Figure 8.3). Also included with these in-wheel motor systems are braking components and the motor drive electronics. Figure 8.4 shows a vehicle with four installed in-wheel motors.

At present, there are several areas where it appears that in-wheel motors can provide advantages. Conventional vehicles implement functions like traction and stability control by slowing down the wheel that is spinning faster than it should. But that approach is rather slow to respond and is limited to applying retarding force. To unlock a skidding tire, it would be preferable to apply some driving torque. With in-wheel motors, you can do that. You can deliver precisely controlled braking or motoring torque on a millisecond timescale and, thereby, greatly improve traction and stability control, reduce stopping distances, and enhance drivability and safety.

In-wheel motors also allow torque vectoring, the application of different torques to different wheels that can improve handling markedly. In a car that uses an in-wheel motor system, this hardware ability comes essentially free, requiring only the right software. As a result, the vehicle can ride through corners as if on rails. In addition, the vehicle would be nimble in city traffic and stable at high speeds.

FIGURE 8.3 In-wheel motor system includes braking components and the motor drive electronics.

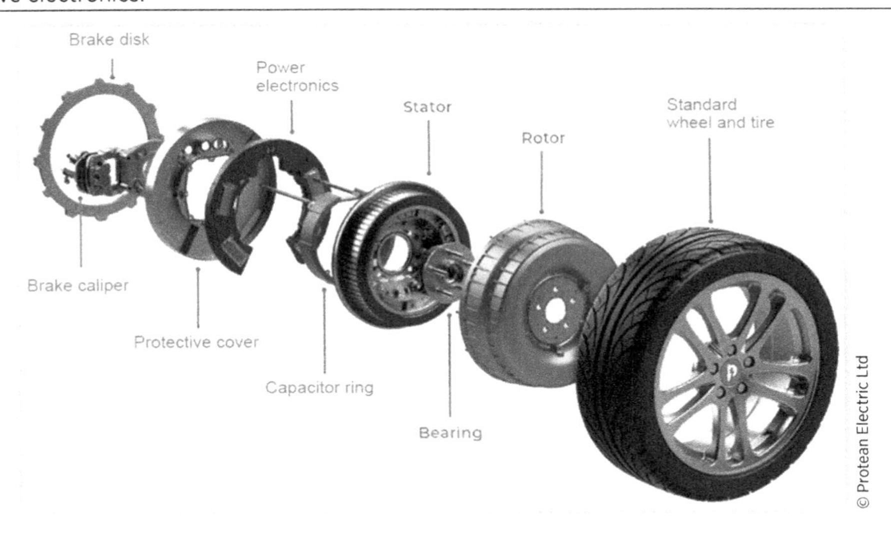

Today's in-wheel technology developers, such as Protean Electric, claim they have overcome or are close to overcoming the challenges of using these motors:

- Cost
- Additional mass
- Road shocks

FIGURE 8.4 Vehicle with four installed in-wheel motors.

© Protean Electric Ltd

Motor control software is also a key design challenge. It must make decisions about what torque to demand from each motor at each instant based on the vehicle's condition and the driver's commands. Under normal circumstances, running two motors instead of just one is straightforward. But if a fault occurs in one motor, the controller needs to prevent a dangerous asymmetry from developing that would cause the vehicle to pull to one side in an uncontrollable way.

Protean Electric's response to this safety issue was to include two completely independent systems that can detect a fault and equalize torque across two or more motors. Vehicles must meet ISO 26262, a safety standard that calls for proving that a hazardous technical failure will be extraordinarily unlikely.

Protean's motors fit behind the wheels of a vehicle, so they can be used as part of a drive system that does not require a gearbox, differential, or drive shafts. This creates an energy-efficient drivetrain that potentially saves cost, reduces weight, and frees up space on board the vehicle that was previously dedicated to drivetrain components. According to Protean Electric, its in-wheel motors can increase fuel economy by over 30%, depending on the battery size and driving cycle in a hybrid or plug-in hybrid vehicle. It is also capable of enabling torque vectoring by applying individual torque at optimal levels to each wheel to improve vehicle safety and handling.

In-wheel motors, however, have two challenges to overcome:

- Reducing unsprung mass because the weight of a motor will be carried in each powered wheel.
- Protecting against road shocks and heat from braking due to the proximity of the in-wheel motor to the wheels.

In-wheel motors add unsprung weight that is the enemy of handling. Every time the car hits a bump, a pothole, or a speed bump, all that weight has to get started in one direction and then move in the opposite direction as soon as the spring it is attached to reaches the limit of its travel.

Unsprung mass is everything between a car's suspension system and the road. In a conventional vehicle, this includes brakes, bearings, wheels, constant velocity joints (the devices on the ends of the drive axles that let them transmit power at an angle), and tires. Keeping unsprung mass to a minimum is a good thing. Reducing it improves ride quality for the driver and passengers and makes it easier for the suspension to keep the tires in contact with the road [11].

8.9 Cooling It

Researchers at the Fraunhofer Institute for Chemical Technology (ICT) are working together with the Karlsruhe Institute of Technology (KIT) to develop a new cooling concept that will enable polymers to be used as EV electric motor housing materials, thereby reducing the weight of the motor and thus, the EV itself. The new cooling concept could also significantly increase the power density and efficiency of the motor, compared with the current state-of-the-art [12].

Fraunhofer researchers see three issues that play a particularly important role when it comes to using an electric motor for mobility:

- High power density
- Compact configuration that fits snugly within the electric vehicle
- High levels of efficiency

The research focus of the DEmil (Directly Cooled Electric Motor with Integral Lightweight Housing made of Reactive Injection Molding) is on a novel electric motor for traction applications. The permanent-magnet-excited synchronous machine with single-tooth winding is optimized for a high power density and efficiency. The central part of the motor is a stator composed of 12 segmented individual teeth, which are wound up with a flat wire. The rectangular wire is coiled in several layers and creates a triangular free space between two neighboring teeth. This free space forms an internal cooling channel for the stator winding and is responsible for the high power density and efficiency of the machine due to the spatial proximity and the resulting optimum heat transfer between the heat source (winding) and the heat sink (cooling medium). Thus, higher continuous loads are possible. Furthermore, the resulting low temperature level within the machine makes it possible to use more inexpensive and high-production process-oriented materials in the region of housing parts. Therefore, the thermoset injection molding process will be in the focus for the overmolding of the stator.

8.10 Turbine-Based Hybrid

A concept car, HIPERCAR, will be available as a four-wheel or two-wheel rear drive, full-bodied car. There are two motors, one at each rear wheel for rear drive, or four electric motors, one at each wheel for all-wheel drive.

This hybrid EV features a 750 V, 42 kWh, or 56 kWh, lithium-ion-cooled and -heated battery pack that you can charge, when required, by a 35 kW micro-turbine range extender, meaning the car can run from fuel available at your local gas station. This negates any range anxiety issues and makes the vehicle independent of any charging infrastructure.

The MITRE (MIcro-Turbine Range Extender) is much lighter, smaller, and more efficient than conventional gasoline engines in cars, though they tend to produce very high temperature exhaust that might be difficult to manage [13].

The MITRE prototype system comes in two power outputs, 23 and 47 bhp, and is about 40% smaller and, at 50 kg, about 50% lighter than an equivalent piston engine (Figure 8.5). The more powerful unit also has a thermal efficiency of about 30%, which matches the best piston engines, and both versions have very low emission outputs. Adding a larger heat exchanger can improve the unit's thermal efficiency to 35%, which ranks the unit alongside high-performance racing engines.

Engineering director Nick Carpenter believes turbine engines are the most efficient answer to improving EV range. "There have been various attempts to introduce turbine engines into passenger cars, but those engines were directly driving the wheels," he says. "Electric vehicle sales are gaining traction now, but people who travel long distances still need more range, and this is where the range extender still provides the best answer."

FIGURE 8.5 Micro-turbine is said to be 50% lighter than an equivalent piston engine.

© Delta Motorsport Ltd

Carpenter postulated that adding more batteries to a car to increase its range has its limits due to the weight of battery cells, but a turbine range extender can drastically increase an EV's range for a tiny offset of emissions.

Carpenter reported that production makes up 10% of battery costs, meaning it's the physical materials that make up 90% of costs. It's for this reason he believes range extenders are the cost-effective answer for now. "I think there'll always be a place for range extenders," says Carpenter. "Even if the technology takes off and we have batteries that can offer 150 miles of motorway driving, it still won't be enough for long-distance drivers."

References

1. Morris, D.Z., "Tesla Veteran Explains How Electric Motors Crush Gas Engines," Fortune, November 17, 2015.
2. Tang, Z., "Overcoming the Challenges of Hybrid/Electric Vehicle Traction Motor Design," Ozen Engineering, July 26, 2015, 1-9.
3. Blue Weave Consulting, "Global EV Motor Report," Blue Weave Consulting, March 16, 2019.
4. Rind, S., Ren, Y., Hu, Y., Wang, J. et al., "Configurations and Control of Traction Motors for Electric Vehicles: A Review," Chinese Journal of Electrical Engineering 3, no. 3 (2017): 1-17.
5. Adams, E., "The Secrets of Electric Cars and Their Motors: It's Not All About the Battery, Folks," The Drive, January 9, 2018, 1-4.
6. Murphy, J., "What's the Difference between AC Induction, Permanent Magnet and Servomotor Technologies," Machine Design, April 1, 2012.

7. Rathi, A., "Audi Won't Say Why the E-Tron Has a Big Battery But Low Range," Quartz, April 12, 2019.
8. Jenkins, J., "A Closer Look at Losses in EV Motors," Charged, March 6, 2019.
9. Wikipedia, "DC-AC Inverter."
10. Moreels, D., "Axial-Flux Motors and Generators Shrink Size and Weight," powerelectronics.com, July 13, 2018, 1-4.
11. Davis, S., "In-Wheel Motor Systems Will Propel EV Performance," powerelectronics.com, July 20, 2018.
12. Fraunhofer Institute for Chemical Technology ICT, "Directly-Cooled Lighter-Weight EV Motor Made with Polymer Housing," Fraunhofer Institute, February 2, 2019, 1-4.
13. Davis, S., "Hybrid Electric Sports Car to Use Micro Turbine Range Extender," powerelectronics.com, September 8, 2017, 1-4.

CHAPTER 9

ECU Circuit Protection

Increased electronic content resulted in more electronic control units (ECUs), particularly with the increase of installed ADAS units [1]. All ECUs are vulnerable to electrical hazards that can insert errors into signal lines or on power supply rails that affect EV operation. To ensure safe, efficient, and reliable EV operation, power management protects the EV's ECUs from these electrical hazards, which include:

- Electromagnetic interference (EMI)
- Transients
- Electrostatic discharge (ESD)
- Power malfunctions

9.1 Electromagnetic Interference

In automotive electronic systems, EMI can adversely affect the performance of internal ICs, as well as other electronic vehicle components that are in proximity. A vehicle might generate EMI externally, causing interference with neighboring electronic systems. Also, external systems can generate EMI that interferes with automotive system operation [2–5].

Related to EMI is electromagnetic compatibility (EMC) defined as a state that exists when all devices in a system can function without error in their intended electromagnetic environment. There are three essential elements to EMC:

1. Source of electromagnetic energy
2. Receptor (or victim) that cannot function properly due to the electromagnetic energy
3. Path between them that couples the energy from the source to the receptor.

Each of these three elements must be present, although they may not be readily identified in every situation. EMC problems are generally solved by identifying at least two of these elements and eliminating (or attenuating) one of them.

One way to minimize EMI in ECUs is to partition circuits into EMC critical and noncritical areas [6]. You can leave noncritical circuits alone. When it comes to critical or sensitive regions, you can prevent radiated EMI by adding filters as necessary at the appropriate interfaces. After you partition the circuits, you can start layout design.

The EMC critical areas are those areas that contain sources of radiation or are susceptible to radiation. These areas may include circuits containing high-frequency circuits, low-level analog circuits, and high-speed logic, including microprocessor circuits. Noncritical EMC areas are those that are not likely to radiate signals or be susceptible to radiation. This could include linear power supplies (not switch-mode power supplies) and slow speed digital and analog circuits.

By isolating the EMC critical areas, it is possible to add the relevant measures both at the initial stages of the design or possibly later. Having an interface provides the possibility for optimizing the overall performance to meet its EMC test. This may result in the addition of further filtering, screening, and so on, or it may even enable cost reductions if some of the measures are not required.

9.2 CISPR

The *Comité International Spécial des Perturbations Radioélectriques* (CISPR); English: International Special Committee on Radio Interference) was founded in 1934 to set standards for controlling electromagnetic interference in electrical and electronic devices [7].

CISPR consists of six technical and one management subcommittees, each responsible for a different area, defined as:

- A: Radio-interference measurements and statistical methods
- B: Interference relating to industrial, scientific, and medical radiofrequency apparatus to other (heavy) industrial equipment, overhead power lines, high-voltage equipment, and electric traction
- D: Electromagnetic disturbances related to electric/electronic equipment on vehicles and internal combustion engine powered devices
- H: Limits for the protection of radiofrequencies
- I: EMC of information technology equipment, multimedia equipment, and receivers

CISPR's standards cover the measurement of radiated and conducted interference, as well as immunity for some products. These standards include:

CISPR 1:

- Industrial, scientific, and medical equipment
- Radiofrequency disturbance characteristics
- Limits and methods of measurement

CISPR 12:

- Vehicles, boats, and internal combustion engines
- Radio disturbance characteristics
- Limits and methods of measurement for the protection of off-board receivers

CISPR 25:

- Vehicles, boats, and internal combustion engines
- Radio disturbance characteristics
- Limits and methods of measurement for the protection of onboard receivers

IEC 61000-6-4:

- EMC
- Part 6-4: Generic standards
- Emission standard for industrial environments

For automotive electronics systems, CISPR 25 is the leading benchmark and requirement for body electronics. Therefore, electronic suppliers are now focused on proving that their devices can meet CISPR 25. A CISPR guide for the application of its standards is available in the EMC zone of the IEC website.

EMI testing and compliance is tested according to the test procedure defined in ANSI C63.4-2009 "Methods of Measurement of Radio-Noise Emissions from Low-Voltage Electrical and Electronic Equipment in the Range of 9 kHz to 40 GHz." This ANSI Standard does not include either generic or specific product-related limits on conducted and radiated emissions. These limits are specified in the Federal Communications Commission (FCC) and CISPR documents. It is worth noting that testing is done with the entire system, not just the power module, especially with embedded power modules. With external power supplies (as in standalone power adapters), the entire system needs to be tested, even if the power adapter complies with the regulations.

CISPR 25:2016 contains limits and procedures for the measurement of radio disturbances in the frequency range of 150 kHz to 2,500 MHz. The standard applies to any electronic/electrical component intended for use in vehicles, trailers, and devices. The limits are intended to provide protection for receivers installed in a vehicle from disturbances produced by components or modules in the same vehicle. The method and limits for a complete vehicle (whether connected to the power mains for charging purposes or not) are in Clause 5, and the methods and limits for components/modules are in Clause 6.

Only a complete vehicle test can be used to determine the component compatibility with respect to a vehicle's limit. The receiver types to be protected include broadcast receivers (sound and television), land mobile radio, radio telephone, amateur, citizens' radio, satellite navigation (GPS, etc.), Wi-Fi, and Bluetooth. According to this standard, a vehicle is a machine that is self-propelled by an internal combustion engine, electric means, or both. Annex A provides guidance in determining whether this standard is applicable to equipment. This fourth edition (CISPR 25:2016) cancels and replaces the third edition published in 2008. This edition constitutes a technical revision that includes significant technical changes with respect to the previous edition: inclusion of charging mode for EV and plug-in hybrid electric vehicles (PHEV) [8].

9.3 Grounding

Circuit grounding can affect EMC performance. Poor grounding can lead to ground loops that can cause signals to be radiated, or picked up within the unit, resulting in poor EMC performance.

To help ensure that the grounding system works satisfactorily, you must realize its function. Ground is a path that enables a current to return to its source. It should obviously have a low impedance and should also be direct. Any loops or deviations may give rise to spurious effects that can cause EMC problems.

Planning proper grounding is not trivial. It is more challenging than it appears but essential for a good EMC performance. Wire lengths and p.c. board traces must be kept to a minimum because, in frequencies, more than a few kilohertz the impedance is dominated by inductance, and lengths of a few inches make a difference, even at low frequencies.

To overcome grounding problems, use thick wires, if possible, and use ground planes on p.c. boards. Critical tracks must be run above the ground plane and should be routed, so they do not encounter any breaks in the ground plane. Sometimes it is necessary to have a slot or break in a ground plane and, if this occurs, a critical trace must be routed over the plane, even if it makes it slightly longer.

These and other approaches can be adopted to ensure that the grounding system is able to minimize EMC. Give considerable thought to a grounding protocol, it probably won't be easy to change later.

9.4 Ground Loops

All current that flows into an IC flows back out again, so it is important to use short connections to the IC, including keeping bypass capacitors close to the IC. Often forgotten is the path that the ground current must take to get back to its source. In an ideal situation, a layer of the p.c. board should be dedicated to ground, and its path is not much longer than a via. However, some board layouts have cutouts in ground planes that can force ground currents to take a long path from the chip back to the power source. The ground current taking this path could act as an antenna to transmit or receive noise.

When a magnetic field is present, a loop of conductive material acts as an antenna and converts the magnetic field into a current flowing around the loop. The strength of the current is proportional to the area of the enclosed loop. Eliminate loops as much as possible and keep any required enclosed areas as small as possible. An example of a loop that might exist is when there is a differential data signal. A loop can form between the transmitter and the receiver with the differential lines.

Another common loop is when two subsystems share a circuit, perhaps a display and an ECU that drives the display. There is a common ground (GND) connection in the chassis of the vehicle-a connection to this GND at the display end and at the ECU end of the system. When the video signal is connected to the display with its own ground wire, it can create one huge loop within the ground plane. In some cases, a loop like this is unavoidable. However, by introducing an inductor or a ferrite bead in the connection to ground, a DC loop can still exist, but, from an RF emissions standpoint, the loop is broken.

When looking to solve EMI issues reduce, the effects of dv/dt (change in voltage with respect to time) and di/dt (change in current with respect to time) wherever possible. In this context, switch-mode DC–DC converters may seem completely harmless until it is realized that they don't convert directly from DC to DC. Rather, they go from DC to AC to DC. Hence, the AC in the middle has the potential to cause EMI problems.

9.5 Shielding

Radiated EMI is often emitted by unintended sources that act as antennae, such as internal wires and components. Enclosing an ECU circuit section in a conductive, grounded screen can minimize the radiated EMI. The conductive wall will provide a barrier to radiation, thereby improving both emissions and susceptibility to external EMI.

To minimize emissions, put a shield around the offending portion of the circuit. It may still emit energy, but good shielding can capture the emissions and send them to ground before they escape from the system. Shielding can take a variety of forms. It might be as simple as enclosing a system in a conductive case, or it could involve fashioning small custom metal enclosures that are soldered over emission sources.

It is also possible to spray the inside of enclosures with conductive paint, although this level of shielding will probably not be as good as if a fully conductive metal case. Where high EMC performance is required, select a case where the continuity of the shield is not breached. The case should ideally be made of as few elements as possible. At each joint there is a possibility of radiation passing through. Where joints do occur, they must be as tight as possible and have good continuity between them. For an ECU enclosure consisting of two halves, use a conductive EMI gasket between the two halves to ensure that the two halves of the enclosure have complete electrical contact all the way around.

9.6 Switch-Mode Power Supplies EMI

By virtue of their switching characteristics, switch-mode power supplies may generate EMI at multiple frequencies, or harmonics. The internal switched AC voltage is usually a square wave that you can represent as a Fourier series consisting of the algebraic sum of harmonically related frequencies. These multiple frequency signals are the source of conducted and radiated emissions that can interfere with circuits in nearby equipment that is susceptible to these frequencies [9, 10].

EMI generation by switch-mode power supplies is subject to FCC and CISPR regulations.

The U.S. agency responsible for regulating communications is the FCC. Control of EMI is outlined in Part 15 of the FCC rules and regulations. FCC rules decree that any spurious signal greater than 10 kHz are subject to these regulations. Radiated emissions, that is, those radiated and coupled through the air, must be controlled between 30 and 1,000 MHz. Conducted emissions, that is, those RF signals contained within the AC power bus, must be controlled in the frequency band between 0.45 and 30 MHz.

The FCC further categorizes digital electronic equipment into:

- Class A (designated for use in a commercial, industrial, or business environment excluding residential use by the general public)
- Class B (designated for use in a residential environment notwithstanding use in commercial, business, and industrial environments). Examples of Class B devices are personal computers, calculators, and similar devices for use by the general public. Emission standards are more restrictive for Class B devices because they are more likely to be located close to other electronic devices used in the home.

A standard widely used in the European Community is CISPR 22 (International Special Committee on Radio Interference), "Information Technology Equipment-Radio Disturbance Characteristics-Limits and Methods of Measurement," issued in 1997. Unlike the FCC that regulates electromagnetic interference in the United States, CISPR is a standards organization without regulatory authority. However, CISPR standards have been adopted for use by most members of the European Community.

CISPR 22 also differentiates between Class A and Class B devices and establishes conducted and radiated emissions for each class. Additionally, CISPR 22 requires certification over the frequency range of 0.15–30 MHz for conducted emissions (the FCC range starts at 0.45 MHz).

9.7 Spread Spectrum

Switching power supplies may generate conducted and radiated interference. One technique employed to reduce the effects of conducted and radiated interference is spread spectrum frequency modulation (SSFM) [11]. This technique spreads the noise over a wider band, reducing the peak and average noise at a particular frequency. The key determinants for effective SSFM are the amount of frequency spreading and the modulation rate. For switcher applications, spreading of ±10% is typical, and the optimal modulation rate is dictated by the modulation profile.

One area where automotive designers are concerned about creating interference is in the AM radio band. Most automobiles have an AM radio that has a very sensitive, high-gain amplifier tunable from 500 kHz to 1.5 MHz. If a component emits a signal within this band, it will probably be audible on the AM radio. Many switching power supplies use switching frequencies within this AM band, which leads to issues in automotive applications. As a result, most automotive switch-mode supplies should use switching frequencies that are above this band, sometimes even 2 MHz or higher. If there is insufficient filtering at either the input or the output of a switching power supply, some of this switching noise may find its way into other subsystems that may be sensitive to the root or subharmonic frequencies.

9.8 Cables EMI

One of the major areas for automotive EMI compliance is the RF radiated emissions from ECU's interconnecting cables and their susceptibility to receiving interference. Cables can form a coupling path for interference in a vehicle. Often these cables need to carry high-frequency signals, possible data, which can present some challenges in terms of improving their EMC/EMI performance.

Any cable will receive and radiate signals, especially when it approaches a quarter wavelength, or an odd multiple thereof, because it forms a resonant circuit. However, EMI can be a problem even when the cable does not approach these lengths.

One approach that can help is to carry signals in a differential format. The signal cables can then be constructed as a twisted pair and could even be screened. In this way the high-frequency signal can be carried, but its susceptibility to radiation and reception is reduced, because anything received will appear on both lines and cancel out. Also, radiation does not occur for the same reason.

One solution is to filter the cables entering and leaving a unit. Although this reduces EMI, it may also degrade the performance of the circuit. If high-speed data needs to be carried, then any sharp edges will be removed by the filters and, in the worst case, the signal may be attenuated to such a degree that the system does not work. Thus, the filter design should provide a careful balance between equipment performance and EMC requirements.

9.9 Transient Protection

Transients are very steep voltage steps that happen with the sudden release of a previously stored energy, either inductive or capacitive. Voltage transients may occur in a predictable manner from controlled switching actions, or randomly induced into a circuit from external sources. Transients can range from a few volts to over several kilovolts. Transient spikes can exist for a very short periods (ms or μs) or occur randomly. Voltage transients do not always start at zero volts or at the beginning of a cycle but can be superimposed onto another voltage [12].

A transient voltage suppressor (TVS) is a voltage-clamping diode with an impedance that depends on the current flowing through it. Under steady-state operating conditions, the device offers a high impedance, so it has no effect on the connected circuit. When a voltage transient occurs, the impedance of the device goes down as the current drawn through the device rises. This clamps the transient voltage. This volt-ampere characteristic is generally time dependent as the large increase in current results in the device dissipating a lot of energy.

When using a TVS diode to protect a signal line, consider its reverse stand-off voltage, that is, the transient voltage you do not want the signal to exceed. For example, select a TVS diode with a reverse stand-off voltage of at least 5 V for a 0–5 V signal range. Another important parameter is breakdown voltage that defines the voltage limit at which the TVS diode will begin conducting significant current to ground and provide transient protection. Usually, the breakdown voltage should be slightly higher than the reverse stand-off voltage, so transient protection begins as soon as possible. A third parameter is the clamping voltage, which is the lowest voltage your system will see during a high-voltage transient. The last parameter to review is the amount of capacitance the TVS diode will add across the signal lines. To ensure that a TVS diode doesn't affect a high-speed signal transmission, choose a low capacitance device.

Varistors, zeners, and TVS diodes are used for transient protection. Devices with higher clamping voltage take longer to dissipate transient energy. TVS diodes handle transients more efficiently. Varistors are nonlinear, two-element devices that drop in resistance as voltage increases so they can be used as transient surge suppressors. A zener diode placed across a signal line will not conduct unless the transient voltage rises above the zener voltage.

Although switching transients occur more frequently, and equipment failures unexpectedly occur often after a time delay, degradation of electronic components within a vehicle is accelerated due to the continual stress caused by switching transients.

ISO 7637-2:2011 specifies test methods and procedures to ensure the compatibility to conducted electrical transients of equipment installed on passenger cars and commercial vehicles. It describes bench tests for both the injection and measurement of transients. It is applicable to all types of road vehicles independent of the propulsion system.

9.10 Electrostatic Discharge

ESD is the rapid, spontaneous transfer of electrostatic charge induced by a high electrostatic field. ESD can change the electrical characteristics of a semiconductor device, degrading or destroying it [13]. ESD may also upset the normal operation of an electronic system, causing equipment malfunction or failure.

When you look at the data sheets for semiconductor devices, you will often see human body model (HBM) and charge device model (CDM) ESD ratings [14]. The HBM rating provides an indication of the transient suppressor's resilience to ESD damage from human handling, while the CDM rating is related to the IC's ability to resist ESD damage during automated manufacturing. A 2,000 V CDM ESD test with a standard pulse of 1 ns will generate a 2–5 A that the semiconductor transient suppressor (TVS) will need to survive. A 2,000 V HBM ESD test with a standard pulse of 150 ns will generate about 1.3 A that the TVS will have to survive.

SAE Standard J1113/13_201502 specifies the test methods and procedures necessary to evaluate electrical components intended for automotive use to the threat of ESDs. It describes test procedures for evaluating electrical components on the bench in the powered mode and for the packaging and handling non-powered mode.

ISO 10605:2008 specifies the ESD test methods necessary to evaluate electronic modules intended for vehicle use. It applies to discharges in the following cases:

- ESD in assembly
- ESD caused by service staff
- ESD caused by occupants

ESD applied to the device under test (DUT) can directly influence it. ESD applied to neighboring parts can couple into supply and signal lines of the DUT in the vehicle and/or directly into the DUT.

ISO 10605:2008 describes test procedures for evaluating both electronic modules on the bench and complete vehicles. It also describes a test procedure that classifies the ESD sensitivity of modules for packaging and handling. ISO 10605:2008 applies to all types of road vehicles regardless of the propulsion system.

9.11 Power Malfunctions

A fuse is an electrical safety device that operates to provide overcurrent protection of an electrical circuit [15]. Its essential component is a metal wire or strip that melts when too much current flows through it, thereby interrupting the current. It is a sacrificial device; once it operates it is an open circuit and it must be replaced. A fuse is an automatic means of removing power from a faulty system. Circuit breakers can be used as an alternative to fuses but have significantly different characteristics.

Internally, a fuse consists of a metal strip or wire fuse element of small cross section compared to the circuit conductors mounted between a pair of electrical terminals and enclosed by a noncombustible housing. The fuse is connected in series to carry all the current passing through the protected circuit. The resistance of the element generates heat due to the current flow. The size and construction of the element is (empirically) determined so that the heat produced for a normal current does not cause the element to attain a high temperature. If too high a current flows, the element rises to a higher temperature and either directly melts or melts a soldered joint within the fuse, opening the circuit.

Fuse selection depends on the characteristics of its load. Semiconductors may need a fast or ultrafast fuse as semiconductor devices heat rapidly when excess current flows. The fastest blowing fuses are designed for the most sensitive electrical equipment, where even a short exposure to an overload current could be very damaging. Normal fast-blow fuses are the most general-purpose types. A time-delay fuse (also known as an anti-surge or slow-blow fuse) is designed to allow a current that is above the rated value of the fuse to flow for a short period of time without the fuse blowing. These types of fuses are used on equipment, such as motors, that can draw larger than normal currents for up to several seconds while coming up to speed.

TABLE 9.1 Fuse characteristics.

Parameter	Description
Rated Current (I_N)	Maximum current that the fuse can continuously conduct without interrupting the circuit.
Blowing Speed	Depends on the current that flows through it and the fuse material. Operating time decreases as the current increases.
Standard Fuse	Open in 1 s at twice its rated current.
Fast-Blow Fuse	Blows in 0.1 s at twice its rated current.
Slow-Blow Fuse	Takes tens of seconds to blow at twice its rated current.
I^2t Rating	Amount of energy let through by the fuse element when it clears the electrical fault.
Melting I^2t	Proportional to the amount of energy required to begin melting the fuse element.
Clearing I^2t	Proportional to the total energy let through by the fuse when clearing a fault.
Breaking Capacity	Maximum current that can safely be interrupted by the fuse. This should be higher than the prospective short-circuit current
Rated Voltage	Must be equal to or, greater than, what would become the open circuit voltage. Rated voltage should be higher than the maximum voltage source it would have to disconnect.

Table 9.1 lists important fuse characteristics.

Fuses are used to protect EV systems exposed to high voltage, such as the battery management system, as well as onboard and off-board chargers. These systems involve voltages up to 1,000 V and current in hundreds of amperes. These fuses must meet AEC-Q200.

AEC-200 qualified fuses are available from several suppliers. Some fuses are AEC-Q200 qualified and ISO/IATF 16949 certified. One type of fuse has a solid body for optimum performance under the hood or in the cabin. Its characteristics are:

- Stable at high temperature and high stress
- Good thermal and mechanical performance
- Operating temperature ranges: −55°C to 150°C

9.12 Resettable Fuse

Bourns® Multifuse® Polymeric Positive Temperature Coefficient Resettable Fuses provide overcurrent protection. It offers greater reliability, longer part life, and these fuses can be located close to the load being protected instead of traditionally locating fuses in a fuse box. They are qualified to AEC-Q200-Rev D, which defines the stress test requirements and reference test conditions for qualification of passive electrical devices in automotive applications. Figure 9.1 shows the Bourns MF-SM030 PPTC resettable fuse.

Bourns' production facility for Multifuse® PPTC resettable fuses is certified to IATF 16949, which emphasizes the development of a process-oriented, quality-management system that provides for continual improvement, defect prevention, and reduction of variation and waste in the supply chain.

FIGURE 9.1 Bourns MF-SM030 PPTC resettable fuse is compliant with AEC-Q200 Rev-C-Stress Test, Packaged per EIA 481-2. Characteristics are V_{max} 60V, I_{MAX} 40A R_{MIN} 0.90 Ω, and R_{MAX} 4.80 Ω.

References

1. STMicroelectronics, "Protection of Automotive Electronics from Electrical Hazards, Guidelines for Design and Component Selection," AN2689 Application Note, October 2012, 1-42.
2. Sauerwald, M., "Ten Tips for Successfully Designing with Automotive EMC/EMI Requirements," *Analog Applications Journal*, AAJ3Q2015 (July 2015).
3. Di Paulo, M., "Automotive EMC," Interference Technology, September 21, 2018.
4. Fenical, G., "A Primer on Automotive EMC for Non-EMC Engineers," Compliance, March 1, 2014.
5. Sauerwald, M., "Meeting Automotive EMC/EMI Requirements," Compliance, June 30, 2017.
6. Crowder, D., "10 Steps to Successful Automotive EMC Testing," Elite Electronic Engineering Inc., January 10, 2017, 1-14.
7. IEC, "CISPR International Special Committee on Radio Interference," IEC.
8. Kang, J., An, J., Bae, J., Lee, J. et al., "A Study of Electro-Magnetic Compatibility about Electric Vehicle's Charging Mode," March 29, 2013.
9. CUI, "Electromagnetic Compatibility Considerations for Switching Power Supplies," CUI, 2013.
10. Hegarty, T., Loke, R., and Pace, D., "Understanding EMI and Mitigating Noise in DC/DC Converters," Texas Instruments, May 11, 2017.
11. Scott, K. and Zimmer, G., "Spread Spectrum Frequency Modulation Reduces EMI," Analog Devices.
12. Texas Instruments, "What Is a Transient Voltage Suppressor (TVS) Diode?," Texas Instruments, March 23, 2016.
13. Wikipedia, "Electrostatic Discharge."
14. Hidyard, B., "ISO 10605 Road Vehicles Test Method for Electrical Disturbances from Electrostatic Discharge," TI SLVA054, July 2018.
15. Wikipedia, "Fuses."

EV Thermal Management

Thermal management is the ability to control the temperature of a system by means of technology based on thermodynamics and heat transfer. Expanded use of electronic content in EVs led to the increased need for thermal management to maximize system performance and reliability by removing high heat flux generated by electronic devices [1].

In an EV the heating of semiconductors poses the biggest need for thermal management. For reliable operation, the semiconductor's junction temperature must fall within its maximum temperature range. Operating at too high a temperature affects system reliability.

10.1 Cooling Methods

Table 10.1 lists the possible cooling methods that can be employed in an EV [2].

TABLE 10.1 Electronic system cooling methods.

Cooling method	Description
Conduction cooling	Heat transfer from a hotter part to a cooler part by direct contact.
Natural convection	Based on the fluid motion caused by the density differences in a fluid due to a temperature difference. The higher the fluid flow rate, the higher is the heat transfer rate.
Forced convection	Using an air mover to blow the air through the surrounding components increases the fluid flow rate which, in-turn, increases the heat transfer rate. Forced convection is up to 10 times more effective than natural convection.
Liquid cooling	Liquids have higher thermal conductivity rates than air, so liquid cooling is more effective. However, due to the possibility of leakage, corrosion, extra weight, and condensation, liquid cooling is usually preferred for applications that involve high power densities.

10.2 Thermal Management and Reliability

Because device performance is significantly affected by temperature, thermal management is a key part of the ECU design process. The useful lifetime of electronic components can be decreased significantly because of high thermal stresses. There is a rule of thumb that, for every 10°C increase in temperature, component life is cut in half. This rule of thumb is based on the Arrhenius equation, which assumes the rate of chemical reaction corresponds to the damage to devices over time. This rule of thumb is an approximation and is only valid for a failure mechanism with a specific combination of activation energy/operating temperature and only if that is the mechanism that leads to failure in a component.

You can use the Arrhenius equation to compare the damage accumulated over time for different operating temperatures:

$$R = A^{\left(\frac{E_A}{kT}\right)} \tag{10.1}$$

where

R is the rate of chemical reactions
A is the constant related to reaction
E_A is the activation energy associated with the reaction
k is the Boltzmann constant ($8.617 \times 10^{-5\,eV/K}$)
T is the absolute temperature

10.3 Cooling the Battery

The EV's battery also has thermal management issues. Heat is generated in the battery pack when charging and discharging current. Also, the internal resistance of the battery cells and interconnections varies during vehicle acceleration and deceleration. Like the motors and power electronics components, the EV's battery is sensitive to operating temperature [3–5].

A battery thermal management system (BTMS) is necessary to prevent temperature extremes, ensure proper battery performance, and achieve the expected life cycle. An effective BTMS keeps cell temperatures within their allowed operating range [6, 7].

As defined by engineers at the U.S. Department of Energy's National Renewable Energy Laboratory (NREL), EV battery pack thermal management is needed for three basic reasons:

1. To ensure the pack operates in the desired temperature range for optimum performance and working life. A typical temperature range is 15°C–35°C.
2. To reduce uneven temperature distribution in the cells. Temperature differences should be less than 3°C–4°C.
3. To eliminate potential hazards related to uncontrolled temperature, for example, thermal runaway.

There are two types of cooling systems for batteries:

- Passive cooling (air-cooled) systems that depend on dissipating heat through natural convection and radiation. They often rely on increasing thermal inertia of the system by adding more thermal mass.
- Active cooling (liquid-cooled) systems that rely on cooling fluid forced through the battery by means of a blower or positive displacement pump. The cooling fluid can be air or liquid.

A cooling matrix is normally employed for cylindrical battery cells. A hollow cooling shell is made with holes for containing the cells. Cooling liquid enters one end of the jacket and is made to exit at the diagonal end. The cooling matrix system is robust and helps maintain temperature limits by surrounding the cell with large thermal mass. It has the disadvantage that it requires space and lowers the pack energy density. It also adds weight not only for the amount of metal in the jacket but also for the liquid body [8].

10.4 Air-Cooled vs. Liquid-Cooled

Higher-performance EVs usually opt for liquid cooling of batteries because the high heat capacity of a liquid allows much more control over heat dissipation. However, pushing a possibly conducting liquid through an electric environment is hazardous because the two must be isolated. This almost always ensures that liquid cooling system costs more than its air-cooled cooled counterpart.

Generally, air-cooled systems are employed for applications that are not heavy duty. Air-cooled techniques are low cost and uncomplicated compared to their liquid-cooled counterpart. The drawback of an air-cooled system is that it's difficult to maintain an even temperature across the pack. Air-cooled systems are identified by blowers and air filters connected to the battery pack.

Cold plates with either serpentine or concentric channels running through it is one of the most cost-effective solutions for liquid cooling. One of its disadvantages is that cold plates tend to add extra weight to the pack. Additionally, they are mostly deployed underneath the cells, which is not the most effective way of heat dissipation.

10.5 Air-Cooled Systems

Phase change materials (PCMs) use both the principles of efficient heat transfer and phase change to shift the heat away from the source. The selection of PCM depends on the application. The paraffin wax used in its normal state (unheated state) is a semi-viscous solid. Once it absorbs heat, it converts into liquid. Capillary action or gravity enables the liquid to go through a heat pipe loop. The PCM then loses the heat in the condenser section located far away from the battery.

Baffle configuration is the simplest of all cooling systems. Air is blown by a fan over batteries lumped together. The air is not confined to channels within the battery pack. It is directed to different parts of the pack through baffles. A disadvantage of this configuration is that, other than changing air speed of the fan, there is little control over the heat dissipation.

Cooling fins dissipate heat from the wide face of the cell where most heat is available for dissipation. A single cooling fin is positioned between two cells. The thin cooling fin has a series of channels running through it. The cooling fins fit into position in-between cooling blocks. This method works for both pouch and rectangular, prismatic Li-ion cells. Its disadvantage is that fins are costly to design. In addition, there are several interfaces for cooling channels that can cause a leak unless adequate sealing measures are taken.

Channel configuration is a technique where air is directed to different corners of a battery pack using different channels. Techniques similar to HVAC ducting are used for regulating the speed and volume flow rate of air in different parts of the pack. It is difficult to balance airflow in the channel, as it is limited to certain inlet speed of air. Channels can also increase space usage, which increases the volume of battery pack.

10.6 Fuel Cell Thermal Management

A critical issue for fuel cell EVs (FCEVs) is adequate thermal management and demand-oriented cooling to avoid safety issues, degradation, and a decrease in efficiency during operation. Proton exchange membrane fuel cell (PEMFC) can only tolerate a small temperature variation. Two factors are critical when designing a cooling system for PEMFCs. First, the nominal operating temperature of a PEMFC is limited to roughly 80°C. This means that the driving force for heat rejection is far less than in an internal combustion engine. Second, nearly the entire waste heat load must be removed by an ancillary cooling system because the exhaust streams contribute little to the heat removal [9].

10.7 ECU Enclosure Thermal Management

Effective ECU thermal management has been challenging because of the trend of decreased size with increased functionality and operating temperatures. ECU thermal management includes the ECU enclosure as well as its internal components [8].

ECU enclosures can use metal matrix composites (MMCs) that consist of a metal or alloy reinforced by a metal matrix. This offers improved strength-to-weight and stiffness-to-weight characteristics in a low-cost light material. There are three types of MMCs:

- **Aluminum-based composites**, such as Al/SiC, are a nonhazardous and lower cost solution than the other metal matrix composite competitor materials.
- **Carbon/metal composites** have the advantage of ease of machinability. However, these materials suffer from low values of thermal conductivity (TC), especially for through-thickness TC.

- **Cu-based composites** are considered as one of the most important materials for contemporary thermal management. This is due to its high TC value.

Thermally conductive polymers could be used as alternatives to metal and MMCs in heat transfer applications. Conventional plastics are very poor thermal conductors, but the addition of fillers or additives in plastics makes heat dissipation possible.

10.8 ECU Semiconductor Thermal Management

Discrete power semiconductors usually require thermal management because they can operate at high power levels. The semiconductors usually employ a heat sink, which is a passive heat exchanger that transfers heat. The heat sink is typically a metallic part attached to a semiconductor that releases energy in the form of heat with the aim of dissipating that heat to a surrounding fluid in order to prevent the device overheating (Figure 10.1).

In most applications, the surrounding fluid is air. The device transfers heat to the heat sink by conduction. The primary mechanism of heat transfer from the heat sink is convection, although radiation also has a minor influence. Heat sinks increase the contact surface area between solid and air, thereby increasing heat transfer [2, 10].

There are many designs for heat sinks, but they typically comprise a base and a number of protrusions attached to this base. The base is the feature that interfaces with the device to be cooled. Heat is conducted through the base into the protrusions.

Although heat sink bases are manufactured to be very flat and smooth, their surfaces are rough at a microscopic level. This can result in few points of contact and many tiny air gaps between the semiconductor and its heat sink. Air has a low TC, resulting in poor conduction of heat from the device to the heat sink. To combat this, you can apply a thermal interface material (TIM) to the base of the heat sink to fill these gaps and provide more conduction paths between device and heat sink.

Heat sinks are usually constructed from copper or aluminum. Copper has a very high TC, which means the rate of heat transfer through copper heat sinks is also very high.

FIGURE 10.1 Typical discrete power semiconductor heat sink application with thermal interface material (TIM).

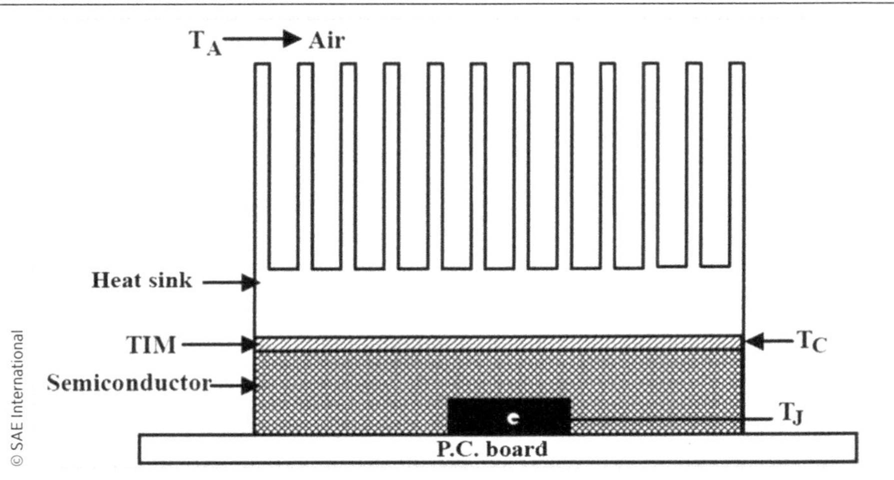

While lower than that of copper, aluminum's TC is still high and has the added benefits of lower cost and lower density, making it useful for EV applications.

Heat sink performance is characterized by its thermal resistance. This parameter can be thought of as the difference in temperature between the air around the heat sink and the device surface in contact with the heat sink per unit of input power. Thermal resistance is denoted by the symbol θ and has the unit °C/W.

In Figure 10.1 the junction temperature (T_J) is the temperature of the hottest part of the device. This is the critical temperature for its correct operation. The case temperature (T_C) is the temperature of the surface of the device, which is in contact with the heat sink assembly. T_C is lower than T_J due to the junction-to-case thermal resistance (θ_{JC}). The performance of the heat sink assembly is defined by the case-to-ambient thermal resistance (θ_{CA}). This is the difference in temperature between the device surface (T_C) and the surrounding air (T_A) for a unit of input power. The increase in T_J over T_A for each watt of thermal power the device generates is the sum of θ_{JC} and θ_{CA}.

10.9 Thermal Interface Materials (TIMs)

There are three types of pastes used as TIMs [11]:

- **Metal-based thermal pastes** use Ag, Cu, or Al as their base. Metal-based thermal pastes offer good TC at a relatively high manufacturing cost. Also, these pastes are electrically conductive, which limits their use.
- **Ceramic-based thermal pastes** are widely used because of their good TC and low cost. These materials are commonly composed of a ceramic powder suspended in a liquid or gelatinous silicone compound. The most commonly used ceramics are beryllium oxide, aluminum nitride, aluminum oxide, zinc oxide, and silicon dioxide.
- **Carbon-based thermal pastes** are electrically nonconductive, so these TIMs are promising as for the future.

One advantage of thermal pastes, or grease, is their lower cost compared to thermal pads or other TIMs. Another advantage is their ability to fill up gaps properly (for rough or irregular surfaces) and reduce the interface resistance between the mating surfaces. However, their disadvantage is in handling this grease, which tends to be messy.

10.10 Thermal Pads

Thermal pads provide a convenient form of TIM. They are usually made from silicone gel technology combined with a thermal medium (usually ceramic). The silicone gel and ceramics are mixed, cast, and cured to a soft, conformable pad with a dough-like texture. In ECUs, thermal pads are typically found between the PCB and the housing base plate. The path of heat removal from an ECU electronics package involves conduction across the interface to the p.c. board surface and then through a thermal pad, into the case or housing (heat sink), and then convection to the environment.

Because thermal pads are not messy, handling of these materials is much easier than thermal grease. They are less likely to be pumped out of the space between the p.c. board and baseplate of the ECU. Thermal pads also act as a vibration damper and protect the electronic components mounted on the PCB. Use of thermal pads could save time by speeding up the assembly process.

Thermal pads are also more stable than phase change products (discussed below) and have higher operating temperatures. However, thermal pads are intended to fill gaps and are not recommended for applications where high mounting pressure is applied on the thermal pad.

Other TIMs include:

- **PCMs** are usually made of suspended particles of high TC, such as fine particles of a metal oxide and a base material. At room temperature, PCMs are solid and easy to handle. At a component's operating temperature these materials change from solid phase to liquid phase. This allows the material to readily conform to both mating surfaces by completely filling up the interfacial air gaps.
- **Thermal conductive insulators** are elastomers that are especially suited for electronics applications because they are electrically nonconductive.
- **Thermal conductive adhesives** are available both in liquid and solid forms (double-sided adhesive tape). Use of these materials eliminates the need for mechanical attachment (i.e., screws, clips, rivets, fasteners, etc.).

References

1. Texas Instruments, "Thermal Management in Automotive," Qpedia, January 14, 2014, 1-3.
2. Molex, "The Need for Thermal Management in Electronic Systems," Molex.
3. Osborne, S., Kopinsky, J., Norton, S., Sutherlan, A. et al., "Automotive Thermal Management Technology," International Council on Clean Transportation, September 21, 2016.
4. Plafke, J., "New Tech Cools Batteries 50-80% More Than Liquid Cooling," ExtremeTech, April 23, 2013.
5. Quesnel, N., "Industry Developments in Thermal Management of Electric Vehicle Batteries," Advanced Thermal Solutions, October 12, 2018.
6. ANSYS, "Battery Thermal Management in Electric Vehicles," ANSYS White Paper, 2011.
7. Li, J. and Zhu, Z., "Battery Thermal Management of Electric Vehicles," Chalmers University, 2014, 1-67.
8. Mallik, S., Ekere, N., Best, C., and Bhatti, R., "Investigation of Thermal Management Materials for Automotive Electronic Control Units," Applied Thermal Engineering 31, no. 2-3, (February 2011): 355-362.
9. Nost, M., Doppler, C., Klell, M., and Trattner, A. Thermal Management of PEM Fuel Cells in Electric Vehicles, Springer Briefs in Applied Sciences and Technology (Cham: Springer International Publishing, 2017).
10. Radian, "What Is a Heat Sink?," Radian.
11. Custom Thermoelectric, "Thermal Interfaces & TIMs," Custom Thermoelectric.

Power Management of ADAS

Automated driver assistance systems (ADAS) is being developed to automate, adapt, and enhance vehicle systems for safety and better driving. Safety features are designed to avoid collisions and accidents by offering technologies that alert the driver to avoid collisions by implementing safeguards and taking over control of the vehicle (Figure 11.1) [1, 2]. This additional electronic content requires effective power management of all its ECUs to ensure the EV functions properly.

11.1 ADAS Basics

ADAS is an informal list that started with a handful of functions provided by car manufacturers and now includes about 20 functions under the ADAS group name. ADAS functions can include sensors, digital processors, cameras, radar, and lasers. These systems can be incorporated in conventional vehicles as well as EVs [3, 4].

As more ADAS functions are added, power requirements will increase. In addition, thermal management will have to be upgraded to keep pace with the power requirements of all the necessary circuits. And these functions will require more ECUs and their attendant circuit protection issues such as EMI and ESD.

Adaptive features may automate lighting, provide adaptive cruise control, automate braking, incorporate GPS/traffic warnings, connect to smartphones, alert driver to other cars or dangers, keep driver in the correct lane, or show what is in blind spots.

There are many forms of ADAS available. Some features are built in cars or are available as an add-on package. However, there are aftermarket solutions available for late model cars.

FIGURE 11.1 ADAS technology will require many sensors and ECUs [2].

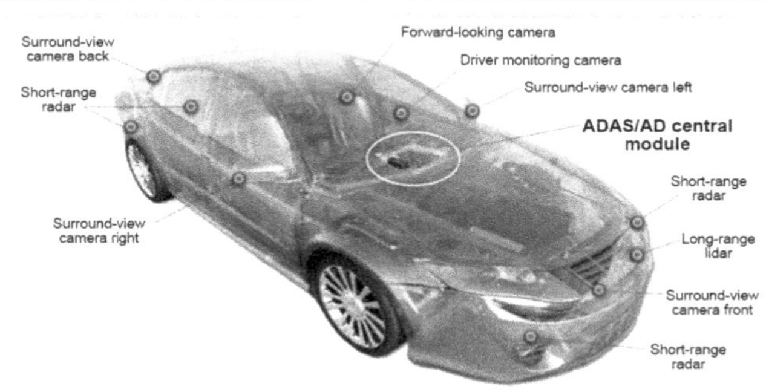

ADAS technology can be based upon vision/camera systems, sensor technology, car data networks, vehicle-to-vehicle (V2V), or vehicle-to-infrastructure systems. These technologies require ECUs with accurate, stable power supplies.

Next-generation ADAS will increasingly leverage wireless network connectivity to offer improved value by using car-to-car and car-to-infrastructure data.

With an average of more than 100 electronic control units (ECUs), modern vehicles are already computer networks on wheels. ADAS systems enhance these vehicles to make them safer and easier to drive, thus more enjoyable.

Some interesting forms of ADAS are special rearview cameras required by NHTSA starting in 2018, face imaging that knows when the driver is nodding off, warnings of local disasters, smart-view mirror, lane departure warning systems, and more.

11.2 National Highway Traffic Safety Administration Rules

The US Department of Transportation's National Highway Traffic Safety Administration (NHTSA) requires all vehicles manufactured on or after May 2018 and under 10,000 pounds to have rearview cameras. The field of view must include a 10-ft by 20-ft zone directly behind the vehicle. The system must also meet other requirements including image size, linger time, response time, durability, and deactivation. The final rule complements action taken by NHSTA to incorporate rear visibility technology into the New Car Assessment Program (NCAP). NHTSA reported that this will increase vehicle safety by significantly reducing the risk of fatalities and serious injuries caused by back-over accidents [5].

American Automobile Association reported that it supports NHTSA's new rule, because rearview cameras can help prevent needless injury and death among our nation's most vulnerable citizens. AAA warned that rearview cameras can be very useful in detecting people and objects behind the vehicle, but there are limitations as with any technology. Drivers will still need to turn and check blind spots behind and to the side of the car while backing up in order to ensure safety.

11.3 ADAS Worldwide

Both the European Union and the United States are mandating that all vehicles be equipped with autonomous emergency braking systems and forward collision warning systems by 2020. A recent McKinsey survey also suggests that car buyers are becoming even more interested in ADAS applications that promote comfort and economy, such as those that assist with parking or monitoring blind spots [2].

Many semiconductor companies—even some that have not traditionally participated in the automotive sector—now offer ADAS products or are developing them. As with any new technology, however, much uncertainty persists about the market, including how consumers will respond to more advanced applications where a computer controls or assists with steering and other critical driving functions.

One factor that could influence ADAS uptake is the rate at which the technology advances. Although semiconductor companies and other players have made important enhancements in recent years, there is much room for improvement.

11.4 ADAS ECUs

ECUs and microcontroller units (MCUs) are essential for most ADAS applications, including autonomous driving. For ADAS to advance, processors need better performance, which could be enabled by multicore architectures and higher frequencies as well as lower power consumption requirements.

To operate properly, ADAS applications require power supplies that meet certain voltage accuracy and load transient requirements. This is necessary to ensure that car battery voltages are properly regulated to sufficiently power ADAS cameras, sensors, and processors in the vehicle's harsh operating environment. The cameras, sensors, and processors require reliable power for successful operation.

With increased in-vehicle processing power comes the need to manage the power supply according to system performance targets [6]. However, given the noisy operating environment of vehicles with their multiple electronic subsystems, balancing power demands with power constraints can be a challenge. To meet these stringent system requirements, automotive power management solutions that deliver precision, flexibility, and small solution size are essential. Thermal constraints, electromagnetic interference (EMI), and heat dissipation are other key issues that must be addressed.

11.5 Commercial ADAS ECU

Ecotrons Automated-driving Control Unit (ACU) is intended to be the core controller of autonomous vehicles or ADAS vehicles (Figure 11.2) [7]. It is based on NVIDIA Xavier processor and can process variant vision sensor data and handle the sensor fusion. It is designed as automotive grade embedded controller; compare to those bulky industry computers. Furthermore, it includes an ASIL-D rated microcontroller, like NXP MPC5744. Combining the super computation power of Xavier and an ASIL-D micro together, Ecotrons ACU offers the production intent autonomous vehicle controllers, which could meet the requirements of both high-performance computation and functional safety, the biggest concern of AVs.

Ecotrons ACU can read data from variant sensors, for example, multiple cameras, radar, LiDAR, and the cloud data; it has enough ports for six cameras, Ethernet input from lidar, CAN bus, Flexray for radar, and other data steam. It can be integrated with a V2X data system and 5G data network. It can be used for L2 to L4 level autonomous vehicles.

FIGURE 11.2 Ecotrons ADAS ECU (ACU).

11.6 Balancing ADAS Power Requirements

Managing the electrical and power considerations of vehicle subsystems requires a delicate balance. Processors, memories, displays, and other components need well-regulated voltages at various current levels. The regulators, in turn, must be efficient in order to deliver the power needed to run these critical circuits without too much heat dissipation. When there are multiple power rails in play, that's when things get really complicated because there are so many more voltage and current spikes to manage. Certain voltage rails in a car have specific voltage accuracy requirements. For example, to guarantee performance levels, system-on-chip (SoC) cores generally have a specified voltage tolerance. And processor performance can't be guaranteed when its operating voltage is out of spec.

There is also the car's electrical and thermal environment to consider. DC rails in cars are noisy. There are large and sudden drops when the car is started under various temperature situations, such as cold cranking, warm cranking, or need to mitigate. Cars may have RF electrical noise from both internal and external sources, causing EMI that can hamper performance in various vehicle subsystems.

Today's vehicles contain a host of electrical subsystems from automotive networking systems to safety systems that are all in proximity in a very confined space. Outside, everything from mobile phones to transmission towers emit noise that can affect the car's performance. Automotive OEMs must ensure that electronic systems do not emit excessive EMI and that they are immune to noise from other subsystems. (CISPR 25 from the International Special Committee on Radio Interference provides a standard for conducted and radiated emissions in vehicles.)

There is also the possibility of load transients like those that occur when the processor suddenly gets an increased power demand and draws more current. For instance, the processor could be in standby mode at one moment, consuming about one-third of its peak power. Then, when the processor is called to action, it could draw the full amount of its current. In this scenario, the switch mode power supply's output voltage would temporarily dip and then bounce around before settling in at its target voltage. The ability to deal with these load transients requires a well-designed power converter to manage the output voltage swing and prevent that swing from impacting system performance.

Texas Instruments points out that automobile safety depends on, among other things, a set of electronics-based technologies designed to aid in safe vehicle operation [9]. ADAS innovations help prevent accidents by keeping cars at safe distances from each other, alerting drivers to dangerous conditions, protecting those in the car and on the street from bad driving habits, and performing other safety-related operations. ADAS also provides functions that will serve as important elements of computer-controlled autonomous operation in the future. If self-driving cars promise to free drivers so they can use their time more effectively during commutes and longer trips; ADAS features will help minimize collision repairs, prevent injuries, and save lives.

Automakers rely on leading semiconductor suppliers for a range of advanced integrated circuit (IC) technologies that can accurately and reliably support a variety of external sensors, communicate among the car's different systems, and provide high-performance, heterogeneous processing for the computer vision and decision-making necessary for next-generation ADAS and automated driving systems.

Nvidia says its NVIDIA DRIVE™ platform takes driver assistance to the next level [9]. The platform employs deep learning and software libraries, frameworks, and source packages that developers and researchers can use to optimize, validate, and deploy their work. This gives developers a powerful foundation for building applications that leverage computationally intensive algorithms for object detection, map localization, and path planning.

NVIDIA's AI solutions allows ADAS to discern a police car from a taxi, an ambulance from a delivery truck, or a parked car from one that is about to pull out into traffic. It can even extend this capability to identify everything from cyclists in the bike lane to absent-minded pedestrians.

11.7 Testing ADAS

Industry estimates suggest that test drives of millions, or even billions, of kilometers would be required for the validation of ADAS functions—most of this in traffic on the public highway. If other aspects are also considered, such as the risk to other road users and the reproducibility of the tests, then it becomes clear that this test scope is practically unachievable in real traffic environments using test and prototype vehicles. Consequently, it is necessary to move the tests into the laboratory, that is to say, into a virtual environment that is suitable for the system to be tested. At the same time, it would be wrong to do away with real tests on the public highway altogether, because the simulation models used in the laboratory can only provide an approximation to the real world.

Vector Software described some steps and methodologies that may be involved in transferring the tests from the road to the laboratory [10]. To integrate an actual ECU and, possibly, also the actual sensors in a simulated vehicle environment (hardware-in-the-loop), it is necessary to establish an electrical connection.

One of Vector's ADAS test environments consist of a virtual test vehicle and a virtual test driver, driving on a simulated, straight test track. The driver drives for 20 s along the flat road without making any steering maneuvers. He drives at an initial speed of 26 m/s (~93 km/h); the accelerator pedal is depressed by 50%.

At the start of the test drive, another vehicle is present as an obstacle outside of the test vehicle's radar range. The vehicle that constitutes this obstacle is initially stationary and then accelerates with the accelerator pedal fully depressed as soon as the test vehicle first reaches the speed of 0 m/s. The "obstacle vehicle" drives in the same direction as the test vehicle.

11.8 Mobileye

Mobileye employs a single-lensed camera to support ADAS and, eventually, autonomous vehicles [11, 12]. This mono-camera can identify shapes, like vehicles and pedestrians, as well as textures like lane markings and traffic sign text. ADAS functions provided by Mobileye include:

- Lane departure warning (LDW) alerts the driver of unintended/unindicated lane departure
- Forward collision warning (FCW) indicates that under the current dynamics relative to the vehicle ahead, a collision is imminent
- Automatic emergency braking (AEB) identifies the imminent collision and brakes without any driver intervention
- Adaptive cruise control (ACC) automatically adjusts the host vehicle speed from its preset value (as in standard cruise control) in case of a slower vehicle in its path

- Lane keeping assist (LKA)
- Lane centering (LC)
- Traffic jam assist (TJA) a combination of both ACC and LC under traffic jam conditions
- Traffic sign recognition (TSR)
- Intelligent high-beam control (IHC)

Mobileye supports ADAS functions with a single camera mounted on the windshield, processed by a single EyeQ® chip. Mobileye employs proprietary computation cores (known as accelerators) optimized for a wide variety of computer vision, signal processing, and machine learning tasks, including deep neural networks. These accelerator cores address the needs of the ADAS and autonomous driving markets. Each EyeQ® chip features heterogeneous, fully programmable accelerators, with each accelerator type optimized for its own family of algorithms. This diversity of accelerator architectures enables applications to save both computation time and chip power by using the most suitable core for every task.

A future version, EyeQ®5, is expected to enable processing of more than 16 multi-megapixel cameras and other sensors. Its computational power will target 15 trillion operations per second, while drawing only 5–6 W in a typical application.

Mobileye uses its one-of-a-kind artificial vision sensor in three steps.

1. The sensor views the road ahead and identifies:
    - Vehicles
    - Pedestrians
    - Cyclists
    - Lane markings
    - Speed limit signs
2. It gathers this information, tracks and measures repeatedly, including
    a. Distance and relative speed of your vehicle relative to other vehicles and pedestrians
    b. Location of your vehicle relative to the lane markings
3. The system then determines if there is a potential danger and then warns the driver with visual and audible alerts. The visual and audible alerts are uniquely positioned to address some of the main causes of collisions singled out by major automotive safety organizations.

According to a 2018 NHTSA report, estimates of US traffic accident were:

- 6,734,000 police-reported motor vehicle traffic crashes
- 36,560 fatalities
- 2,710,000 people injured
- Less than 1% (33,654) were fatal crashes
- 28% (1,894,000) were injury crashes
- 71% (4,807,000) were property damage-only (PDO) crashes

The estimated 6,734,000 crashes in 2018 represents a 4.4% increase from the 6,453,000 police-reported crashes estimated to have occurred in 2017. This was a statistically significant increase [13].

11.9 Standard Terminology

The AAA (American Automobile Association) has proposed standard terminology for ADAS. Terminology is intended to be simple, specific, and based on system functionality. Table 11.1 lists the AAA ADAS functions [14]:

TABLE 11.1 AAA list of ADAS features.

Adaptive Cruise Control	Controls acceleration and/or braking to maintain a prescribed distance between it and a vehicle in front.
Dynamic Driving Assistance	Controls vehicle acceleration, braking, and steering. SAE standard definition of L2 Autonomous systems outlines this.
Lane Keeping Assistance	Controls steering to maintain vehicle within driving lane. May prevent vehicle from departing lane or continually center vehicle.
Blind Spot Warning	Detects vehicles to rear in adjacent lanes while driving and alerts driver to their presence.
Forward Collision Warning	Detects impending collision while traveling forward and alerts driver.
Lane Departure Warning	Monitors vehicle's position within driving lane and alerts driver as the vehicle approaches or crosses lane markers.
Parking Obstruction Warning	Detects obstructions near the vehicle during parking maneuvers.
Pedestrian Detection	Detects pedestrians in front of vehicle and alerts driver to their presence.
Rear Cross Traffic Warning	Detects vehicles approaching from side and rear of vehicles while traveling in reverse and alerts driver.
Automatic Emergency Steering	Detects potential collision and automatically controls steering to avoid or lessen the severity of impact.
Forward Automatic Emergency Braking	Detects potential collisions while traveling forward and automatically applies brakes to avoid or lessen impact severity.
Reverse Automatic Emergency Braking	Detects potential collision while traveling in reverse and automatically applies brakes to avoid or lessen impact severity.
Fully Automated Parking Assistance	Controls acceleration, braking, steering, and shifting during parking. Parking may be parallel and/or perpendicular.
Remote Parking System	Parks vehicle without driver being physically present in vehicle. Automatically controls acceleration, braking, steering, shifting.
Semiautomated Parking Assistance	Controls steering during parking. Driver is responsible for acceleration, braking, and gear position. Parking may be parallel and/or perpendicular.
Surround View Camera	Cameras located around vehicle to present view of surroundings.
Trailer Assistance System	Assists driver during backing maneuvers with a trailer attached.
System Automatic High Beams	Deactivates or orients headlamp beams automatically based on lighting, surroundings, and traffic.
Driver Monitoring	Monitors driver condition by various means to detect drowsiness or lack of attention.
Night Vision	Aids driver vision at night by projecting enhanced images on instrument cluster or heads-up display.
Tire Pressure Monitoring Systems (TMPS)	Monitors tire pressure that can cause a tire failure and result in serious road accidents.

References

1. Automotive World, “Freescale Collaborates with Neusoft and Green Hills Software to Deliver Development Ecosystem for ADAS Vision Applications,” Automotive World, May 19, 2014.
2. Choi, S., Thalmayr, F., Wee, D., and Weig, F., “Advanced Driver Assistance Systems: Challenges and Opportunities Ahead,” McKinsey & Company, February 2016.
3. Wikipedia, “Advanced Driver-Assistance Systems.”
4. Auto Connected Car News, “ADAS Advanced Driver Assistance Systems,” Auto Connected Car News.
5. Walford, L., “NHSTA Goes Forward with Rear View Camera Requirement,” Auto Connected Car News, April 1, 2014.
6. Tsai, W., Jeen, J., and Parikl, C. “Balancing Power Supply Requirements in ADAS Applications,” Maxim Integrated, June 2018.
7. Ecotrons, “ADAS ECU (ADAS),” Ecotrons.
8. Sagar, R., “Making Cars Safer through Technology Innovation,” Texas Instruments SSZY009A, June 7, 2017, 1-11.
9. NVIDIA, “Advanced Driver-Assistance Systems,” NVIDIA.
10. Skanda, D., Gonzalez, F., Neuffer, J., and Phillip, O., “Combining Forces for ADAS Testing,” Vector Software GMbH, September 6, 2018.
11. Mobileye, “Mobileye: How It Works,” Mobileye, 2018.
12. Mobileye, “The Evolution of EyeQ,” Mobileye, 2018.
13. NHTSA, “Police-Reported Motor Vehicle Traffic Crashes in 2018,” DOT HS 812 860, NHTSA, November 2019.
14. Edmonds, E., “AAA Recommends Common Naming for ADAS Technology,” *AAA Newsroom*.

CHAPTER 12

Power Management of Autonomous EVs

Autonomous vehicles (AVs) add to power requirements because they require multiple ECUs and sensors to support AV functions. High-speed processors are particularly high-power consumers. This adds the need for more thermal management and more circuit protection that involves EMI, ESD, and transient protection. A major challenge will be to design an autonomous/driverless vehicle with the necessary electronic systems that allow duplication of human performance. From a power management standpoint, it will require a primary power source that can support all the required electronic systems as well as providing enough power for the traction motor [1, 2].

12.1 SAE Autonomous Levels

AVs will be controlled by a computer, sensors, and various subsystems [3]. SAE J 3016-2018 describes the levels of driving automation now known as the SAE levels [4]. These levels have reached mainstream and official recognition and are referenced in the US Department of Transportation's Comprehensive Management Plan for Automated Vehicle Initiatives.

These levels include the following acronyms:

- **DDT**: Dynamic driving task, or "all the real-time operational and tactile functions required to operate a vehicle in on-road traffic."
- **ODD**: Operational design domain, or "the specific conditions under which a given driving automation system or feature thereof is designed to function."
- **ADS**: Automated driving system, which encapsulates all the hardware and software collectively capable of performing the entire DDT on a sustained basis.

The degree of automation in these mutually exclusive levels increases as the levels increase, with Level 0 featuring absolutely no automation and Level 5 calling for full automation. The six SAE Levels of Driving Automation are listed in Table 12.1.

Many of the major subsystem components required for an AV are already developed. Figure 12.1 depicts the typical subsystems and sensors for an AV. This vehicle will essentially be a computer on wheels. It will require a learning curve beyond anything experienced to date for electric vehicles. Software and its updates will be a major hurdle for the vehicle manufacturer to overcome.

According to a report by the Energy Information Administration, many studies point to the potential societal and individual benefits that underline the interest in

TABLE 12.1 SAE's six levels of vehicle automation [4].

Level	Automation	Description
0	No Driving Automation	The performance by the driver of the entire DDT. This level is for conventional automobiles.
1	Driver Assistance	Characterized by the sustained and ODD-specific execution of either the lateral or the longitudinal vehicle motion control subtask of the DDT. It does not include the execution of these subtasks simultaneously. It is also expected that the driver performs the remainder of the DDT.
2	Partial Driving Automation	Similar to Level 1, but characterized by both the lateral and longitudinal vehicle motion control subtasks of the DDT with the expectation that the driver completes the object and event detection and response (OEDR) subtask and supervises the driving automation system.
3	Conditional Driving Automation	The sustained and ODD-specific performance by an ADS of the entire DDT with the expectation that the human driver will be ready to respond to a request to intervene when issued by the ADS.
4	High Driving Automation	Sustained and ODD-specific ADS performance of the entire DDT is carried out without any expectation that a user will respond to a request to intervene.
5	Full Driving Automation	Sustained and unconditional performance by an ADS of the entire DDT without any expectation that a user will respond to a request to intervene. Please note that this performance, since it has no conditions to function, is not ODD-specific.

FIGURE 12.1 Typical subsystem components for an autonomous vehicle.

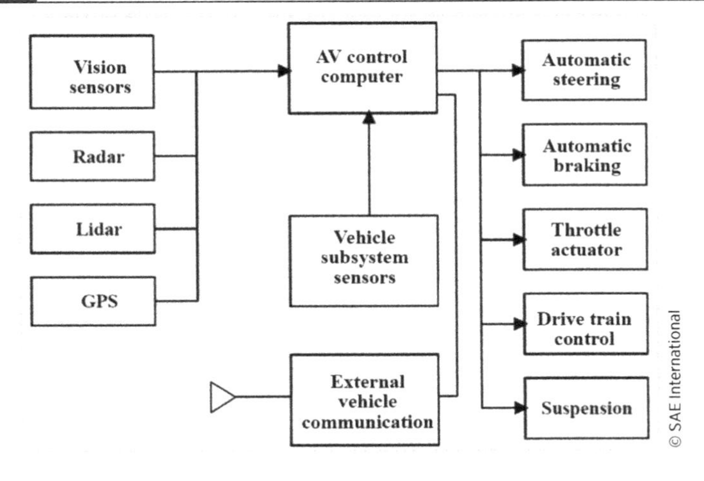

adopting AVs [5]. "These benefits will likely increase as adoption rises, highlighting the importance of consumer and regulatory acceptance, and require increased connectivity to achieve the highest potential. Further, it could take several decades for technologies introduced in newer vehicles to gradually become a large share of the entire vehicle stock."

Accomplishing a self-driving autonomous car will not be "a walk in the park," it will involve solving a broad range of hardware and software technical challenges. And the cost will be substantial.

12.2 Autonomous Vehicle Design Challenges

Autonomous cars require a variety of technologies to detect their surroundings, such as lidar, computer vision, GPS, and odometry. Advanced control systems must interpret sensory information to identify appropriate navigation paths as well as obstacles and relevant signage. These cars must also distinguish between people and different cars on the road [6].

Anil Gupta described "Five challenges in designing a fully autonomous system for driverless cars" [7]. Gupta noted that the real challenge was how to design an autonomous/driverless vehicle system capable of handling vehicle's performance like a human in all possible conditions. "The autonomous vehicle typically is a combination of sensors and actuators, sophisticated algorithms, and powerful processors that execute critical software. There are hundreds of such sensors and actuators that are situated in various parts of the vehicle being driven by a highly sophisticated system."

Gupta said one of the three sensory systems manages a car's internal systems. A significant portion of the design of an AV involves mundane issues such as power management. Several application-specific, unique circuit boards and subsystems are added to a conventional vehicle to provide the functions needed for autonomous operation. Much of the system-level operation involves measuring and managing the power requirements to control power, overall consumption, and thermal dissipation.

Gupta notes that, "challenges are many even today for rolling out the autonomous cars on the road. But so is the determination of scientists, engineers and problem solvers from various disciplines. The collective effort of the industry will definitely make the autonomous car on the road a reality one day, and the benefits will be huge. Not only will it save fuel, encourage efficient transportation, and shared services, but will also help in saving many lives that are regularly lost in road accidents."

12.3 AV Power Struggle

Jack Stewart, author of *Self-Driving Cars Use Crazy Amounts of Power, and It's Becoming a Problem*, said that today's self-drivers don't need extra engines, but they still use terrific amounts of power to run their onboard sensors and do all the calculation needed to analyze the world and make fast driving decisions [8]. And it's becoming a problem.

> *A production car you can buy today, with just cameras and radar, generates something like 6 gigabytes of data every 30 seconds. It's even more for a self-driver, with additional sensors like lidar. All the data needs to be combined, sorted, and turned into a robot-friendly picture of the world, with instructions on how to move through it. That takes huge computing power, which means huge electricity demands. Prototypes use around 2,500 watts, enough to light 40 incandescent light bulbs.*

Additionally, there must be enough power available for the traction motor, which will probably consume more power than all the electronic circuits. This means traction motors must maximize efficiency, which requires minimizing electrical and magnetic losses. Motor losses also generate heat that must be removed from critical components, so effective thermal management of the motor is necessary. This may require new types of motors that provide proper performance at minimum power.

Also, the DC–AC inverter that drives the traction motor must be as efficient as possible, particularly because of the high voltage that is used. This may require wide bandgap semiconductors, such as silicon carbide MOSFETs. Operation at the high power levels means these inverters will require effective thermal management.

Optimizing power management enables an AV to provide the necessary operational performance and maximize its driving range. This requires:

1. High-efficiency ECU power supplies
2. A high-efficiency, high voltage input DC–DC converter that provides power for low voltage electronic loads
3. Electronic systems, sensors, and processors that consume a minimum amount of power
4. Associated batteries or fuel cells that provide enough power for all the electronic circuits as well as the traction motor
5. A traction motor that minimizes power consumption
6. A DC–AC inverter for the traction motor that is power-efficient

12.4 Autonomous Processor

At 2018 CES, Nvidia described a new processor specifically for AVs called Xavier [9]. It has an 8-core CPU (central processing unit) and 512-core GPU (graphic processing unit), a deep learning accelerator, computer vision accelerators, and 8K video processors [10]. The company says it's the most complex system on a chip ever created. It can deliver 30-trillion operations per second, all on a single system on chip (SoC) that consumes 30 W.

The Xavier chip contains an 8-core CPU cluster, GPU with additional inference optimizations, deep learning accelerator, vision accelerator, and a set of multimedia accelerators providing additional support for machine learning. Xavier features a large set of I/O and has been designed for safety and reliability supporting various standards such as functional safety ISO-26262 and ASIL level C.

Sam Abuelsamid of Navigant Research discussed Intel's Challenge to Nvidia for automated driving computing [11]. He said that highly automated driving requires some heavy-duty computing power to make sense of all of the raw data coming from dozens of sensors in self-driving vehicles generating up to 4 TB of raw data every hour. From that data, sophisticated software algorithms must recognize the objects in the field of view, calculate their speed and trajectory, and predict where they are going next. Only then can the system begin to determine what the vehicle should do.

Abuelsamid pointed out the new Intel AV computing platform consists of two of Mobileye's upcoming EyeQ5 sensor processing chips and a new Intel Atom 3xx4 CPU that combine to provide 60% more performance at the same 30 W power consumption as Nvidia's new Xavier chip. As more processing is needed on the way to eventual level 5 AVs that can operate anywhere without human intervention. Cooling and power consumption for early AV development platforms consumed hundreds of watts, or more, before the vehicle even moved.

GM's chief AV program engineer Andrew Farah revealed that the first-generation Chevrolet Bolt EV prototype consumed 3–4 kW to power its complete system. Since AVs in mobility service applications will be expected to operate 20 h or more a day, power consumption could be a problem. The 60-kWh battery in the Bolt would be consumed by a 3-kW automation system running for 20 h before it makes a single trip.

Intel's chief systems architect for autonomous driving systems Jack Weast understands the power efficiency needs of these vehicles, so his team is emphasizing not just raw performance but performance per watt.

> *The Mobileye EyeQ5 processors are each claimed to be capable of performing 24 trillion deep learning operations per second (TOPS) at 10W each. Nvidia claims its new Xavier chip will do 30 TOPS at 30W. The pair of EyeQ5 chips plus the Atom, memory and other related components are projected to match that 30W total while providing more overall performance.*

Abueslsamid points out that, "Intel has opted to use the Atom branding on its CPU for its AV platform. Atom is typically the branding used for the low-power and low performance chips used in inexpensive computers. The AV processor shares the X86 instruction set and low power characteristics with those chips, but also incorporates high-end multi-threading and virtualization technologies from Intel's workstation and server-class Xeon chips. Intel claims this will enable the Atom to run many different tasks currently spread across discrete ECUs in the car while also providing increased security."

Abueslsamid said that Intel's use of two EyeQ5 chips will provide some backup. "The primary function of one chip will be processing the raw sensor signals while the other will handle the fusion calculations that combine all that the sensors see into a single cohesive image of the world around the vehicle. A subset of calculations will actually be carried out in both chips and then compared for consistency before commands are sent over to the Atom chip which will handle path planning and control of the steering, brakes and propulsion. Achieving the same with an Xavier will require two chips, doubling the power consumption to 60 W."

While Intel and Nvidia continue to battle it out, they are also far from the only players. "Established automotive silicon suppliers including NXP, Renesas and even Qualcomm are all hoping to grab what could be a very lucrative business in the next decade and beyond. The great AV computing race has only just begun," said Abueslsamid.

12.5 **AV Power**

A paper entitled "Real-Time Improved Power Management for Autonomous Systems," authored by Mansor et al. [12], said power resources onboard autonomous systems are limited, but power requirements on these systems are increasing due to rapid technology growth. Today's methods for controlling these resources either use expensive and conservative strategies (e.g., reactive control) or employ predefined power schedules that heighten the risk of operation failure in a dynamic environment.

An intelligent power management system (PMS) is required to improve, or maintain, system capability. The strategies proposed in this chapter aim to contribute toward an intelligent PMS. Using optimization methods, an adaptive and flexible PMS capable of constructing the best executable power schedules while satisfying real-time requirements is presented. A three-level optimization strategy is introduced. Due to the feasibility requirement of the solutions produced, the first level uses a constraint satisfaction approach. Then, the solution is quickly improved using a local search algorithm and, next, a global search algorithm is used in the remaining execution time to explore the possibility of further improvement in the solution. The efficiency of the last two levels is enhanced by use of convex programming techniques. Using a case study, they demonstrated that the proposed PMS is capable of rapidly producing a feasible solution and, subsequently, optimizing this solution to provide an improved solution. The proposed PMS can adapt to a dynamic environment by coping with any change in problem description and problem constraints and constructing a new best executable solution while satisfying real time requirements.

Another consideration is the car's maintainability. If the system integrator is getting hardware from several different suppliers, will the circuits be packaged in a similar way with common components? And what kind of test equipment is required to maintain this type of car? Eventually, an established car manufacturer will have to work with the autonomous car company to test its systems.

12.6 Testing Autonomous Cars

If predictions are correct, we can expect to see commercial AVs on our streets within 5–10 years (or less). There are already a small number of these prototype vehicles from Google and its spin-off, Waymo. As of June 2016, Google had road tested their fleet of vehicles, in autonomous mode, a total of 1,725,911 miles. Based on Google's own accident reports, their test cars have been involved in 14 collisions, 13 of which human drivers were at fault. It was not until 2016 that the car's software caused a crash.

Road tests are not normally used on mass-produced conventional vehicles. A few of these vehicles are put through a well-defined, standardized crash test using dummies and sensors. Some cars undergo rollover tests that are performed with a few well-defined steering maneuvers. The results are easily measured to the satisfaction of vehicle buyers, government regulators, and insurance companies. Manufacturers usually run a series of short tests to check each car before they are shipped.

Mass-produced AVs are a different story; manufacturers can run crash tests and rollover tests, but a road test isn't practical or cost-effective for every car that comes off the production line. Instead, AVs will require extensive test procedures to ensure safe and reliable operation. For best results these test procedures should be developed in parallel with the design stage, so they are ready when production starts.

12.7 Static Tests

Two types of tests may be used on autonomous cars: static and dynamic. The static tests involve a check of all the operating voltages. Dynamic tests are related to the software and operation of the sensors working in a real-world environment. It is probably possible to perform the static tests using a computer program that presents the data.

Static tests check the power management of a parked vehicle. This could be done at high and low temperatures of operation. One of the static tests for these vehicles will be to check all the power supplies employed in a vehicle. This could involve "margining," that is, varying power supply voltages of 5%–10% to check proper operation of all systems in the vehicle. This will be necessary as the power source (batteries or fuel cell) age and are recharged multiple times. This would identify any circuits that will not function properly if their power input changes.

Another power-oriented test is to check the voltage applied to each ECU, of which there are quite few as indicated by the list in Table 12.2 of suppliers of these systems. Some of these systems will be purchased while others are produced in-house. It is probable that voltage and current requirements for these systems vary from unit to unit. Input power errors are typically a problem in complex electronic systems. Ideally, each of these make-or-buy systems would be housed in an ECU with similar power requirements.

Another static test would be for the high voltage traction battery. One test would be with the battery management system (BMS), whose job it is to monitor the battery pack output [2]. This includes checking cells connected in series to ensure all cells have about the same output voltage. If not, the cells should be balanced with a balancing circuit that is usually the active type found in a BMS.

Another thing to test is the ability to charge the battery, which could be from an onboard charger or off-board type. There would have to be some way to determine how long it would take to charge battery for a given power input. Today, most EV owners want to want to charge their batteries as quickly as possible, which requires adequate and substantial charging power.

Additionally, there should be a test of the power applied to each system. In many cases the power must be applied sequentially and also removed in the appropriate order. High-performance processing devices, such as FPGAs, ASICs, PLDs, DSPs, ADCs, and microcontrollers, require multiple voltage rails just to power their own internal circuitry, such as the core, memory, and I/O. These applications demand very specific voltage rail power-up and power-down sequencing in order to guarantee reliable operation, better efficiency, and overall system health. If this is not done in the right sequence some systems may not operate properly or even be damaged.

Most semiconductors will have to be checked to ensure they are powered with the correct voltage, particularly the DC–DC converters and analog devices used with the various sensors. Particularly, the operating temperature of all the power semiconductors should be checked. Every place where semiconductors are employed should be monitored.

TABLE 12.2 Companies that provide autonomous vehicle systems.

Company	Location	Expertise
Mobileye	Israel	Powered by an artificial vision sensor that views the road ahead this system identifies vehicles, cyclists, and pedestrians in the vehicle's path and alerts the driver of potential dangers. Also, it can detect lane markings and speed limit signs. It uses a one-of-a-kind artificial intelligence (AI) vision sensor.
Yogitech	Italy	Semiconductor company specializing in adding safety functions to chips used in robots, self-driving cars, and other autonomous devices.
Itseez	San Francisco, CA	An expert in computer vision (CV) algorithms that include methods for acquiring, processing, analyzing, and understanding images from the real world in order to make informed decisions and automotive actions.
Movidius	San Mateo, CA	Vision processing unit (VPU) for autonomous machines that can see in 3D, understand their surroundings, and navigate accordingly.
Nervana	San Diego, CA	Experts in "deep learning"/AI solutions. Deep learning is part of a broader family of machine learning methods based on learning representations of data.
Arynga	San Diego, CA	Provider of software management solutions for secure wireless over-the-air (OTA) updating of software and firmware in the automotive market.
Altera	Santa Clara, CA	FPGAs and FPGA software tools, configurable embedded SRAM, high-speed transistors, high-speed I/Os, logic blocks.
Wind River	Alameda, CA	Computer vision expert. Provides self-driving cars with the ability to see and carefully interpret surroundings.

Therefore, there should be some way to check the thermal management of all the electronic systems.

Another important consideration is the use of fuses or circuit breakers to protect the electronic systems. There will have to be some way to check these safety features without destroying them.

Besides testing the power infrastructure, the rest of the system must be checked, which has its own set of challenges. For example, various systems from the different manufacturers have their own unique packaging. Does that mean the car manufacturer should repackage every ECU so they are all similar? Or should the car manufacturer use the systems as they are now packaged, which could cause a logistics problem for replacement parts?

The reason for comprehensive testing is that AVs are actually computer systems with a network of diverse sensors. The sensors may come from several different companies, each with its own packaging and software, which complicates testing. Thus, testing of AVs is by no means easy. Rigorous testing will be needed to ensure mass-produced AVs are safe and reliable. People will be skeptical of autonomous performance, so they better work properly.

Proper AV operation will depend on hardware and software testing. Ideally, static hardware testing should uncover problems before they are found in the field. However, with conventional vehicles it usually turns out that the user performs the final tests by notifying the manufacturer if there is a hardware problem. You can envision a more complex hardware situation with an AV. To verify this, all you must do is look at all the product recalls for virtually every conventional vehicle.

Moreover, initial software testing must be thorough to eliminate any "glitches." No matter how thorough the testing it will always be necessary to produce software updates. As an example, you constantly get software updates for your PC. Because AVs can include different types of software for its sensors there may be software conflicts that require updates. Software updates are certain to be an important part of the lifetime of an AV.

Another challenge will be different software programs for the electronic systems. Are they going to be able to communicate with each or the main processor, or will someone redo all the software programs so they are similar?

If there will be testing of the autonomous system from 10 different suppliers, will they be able to communicate with each other? In testing, there will always be arguments between hardware and software people when it comes to deciding who is responsible for a specific problem. It's even worse with an AV because of the complexity caused by the number of different technologies and software that are involved. Therefore, testing can be frustrating and time consuming.

12.8 Dynamic Tests

Mass-produced automated vehicles present a totally different situation than just testing the electronic systems. To ensure they will operate safely, vehicles must be tested for different driving scenarios. For example, they will have to be able to avoid other cars, bicycles, and pedestrians. What happens when the AV sees a red or green traffic light? What if a bus swerves in front of the car? Because of the multitude of possible driving situations, testing of an AV requires monitoring many functions.

There is another complication. There will probably be at least a dozen commercially available AVs. Each one might have its own computer and sensors in a unique configuration. It may be possible to have a standard format for the test in which you enter the specifics of your particular vehicle. On the other hand, test systems might have to be unique for each vehicle.

The various functions performed by AVs became apparent to the US government, so it decided to get involved and established policies for these self-driving vehicles. On September 20, 2016, the US Department of Transportation issued a Federal policy for automated vehicles, laying a path for the safe testing and deployment of new auto technologies. The policy sets a proactive approach to providing safety assurance and facilitating innovation through four key parts, summarized:

- 15-point safety assessment—The Vehicle Performance Guidance for Automated Vehicles for manufacturers, developers, and other organizations includes a 15-point "safety assessment" for the safe design, development, testing and deployment of automated vehicles.
- Model State Policy—This policy section presents a clear distinction between Federal and State responsibilities for regulation of highly automated vehicles and suggests recommended policy areas for states to consider with a goal of generating a consistent national framework for the testing and deployment of highly automated vehicles.
- National Highway Transportation Safety Administration (NHTSA) Current Regulatory Tools—This discussion outlines NHTSA's current regulatory tools that can be used to ensure the safe development of new technologies, such as interpreting current rules to allow for greater flexibility in design and providing limited exemptions to allow for testing of nontraditional vehicle designs in a more timely fashion.
- Modern regulatory tools—This section identifies new regulatory tools and statutory authorities that policymakers may consider in the future to aid the safe and efficient deployment of new lifesaving technologies.

The primary focus of the policy is on highly automated vehicles, or those in which the vehicle can take full control of the driving task in at least some circumstances. Portions of the policy also apply to lower levels of automation, including some of the driver assistance systems already being deployed by automakers.

12.9 Evaluating AVs

Federal policies can define the characteristics of self-driving cars and establish safety requirements, but it does not dictate how these vehicles will be tested and evaluated. Evaluation procedures that can measure the safety and reliability of these driverless cars must develop far beyond existing safety tests. To get an accurate assessment in field tests, such cars would have to be driven millions or even billions of miles to arrive at an acceptable level of certainty; a time-consuming process that would cost tens of millions of dollars.

University of Michigan (U-M) is doing its "homework" to develop accelerated testing for AVs [13]. Figure 12.2 shows a typical AV from Waymo researchers affiliated with the U-M's Connected and Automated Vehicle Center. They developed an accelerated evaluation process that eliminates the many miles of uneventful driving activity to filter out only the potentially dangerous driving situations where an automated vehicle needs to respond, creating a faster and less expensive testing program. This approach can reduce the amount of testing needed by a factor of 300–100,000, so that an automated vehicle driven for 1,000 test miles can yield the equivalent of 300,000–100 million miles

FIGURE 12.2 Typical autonomous vehicle from Waymo.

Sundry Photography / Shutterstock.com

of real-world driving. While more research and development needs to be done to perfect this technique, the accelerated evaluation procedure offers a groundbreaking solution for safe and efficient testing that is crucial to deploying mass-produced automated vehicles.

"Even the most advanced and largest-scale efforts to test automated vehicles today fall woefully short of what is needed to thoroughly test these robotic cars," said Huei Peng, director of Mcity and the Roger L. McCarthy Professor of Mechanical Engineering at U-M.

Essentially, the new accelerated evaluation process breaks down difficult real-world driving situations into components that can be tested or simulated repeatedly, exposing automated vehicles to a condensed set of the most challenging driving situations. In this way just 1,000 miles of testing can yield the equivalent of 300,000-100 million miles of real-world driving.

Yet for consumers to accept driverless vehicles, the researchers say tests will need to prove with 80% confidence that they're 90% safer than human drivers. To get to that confidence level, test vehicles would need to be driven in simulated or real-world settings for 11 billion miles. But it would take nearly a decade of round-the-clock testing to reach just 2 million miles in typical urban conditions.

Beyond that, fully automated, driverless vehicles will require a very different type of validation than the dummies on crash sleds used for today's cars. Even the questions researchers have to ask are more complicated. Instead of, "What happens in a crash," they'll need to measure how well they can prevent one from happening.

If an AV crashes, repair personnel will have to be aware of the high voltage for the batteries and traction motor. These personnel will have to be trained to use safety repair procedures to avoid injuries.

"Test methods for traditionally driven cars are something like having a doctor take a patient's blood pressure or heart rate. Testing for automated vehicles is more like giving someone an IQ test," said Ding Zhao, assistant research scientist in the U-M Department of Mechanical Engineering.

To develop the four-step accelerated approach, the U-M researchers analyzed data from 25.2-million miles of real-world driving collected by two U-M Transportation Research Institute projects—Safety Pilot Model Deployment and Integrated Vehicle-Based Safety Systems. Together they involved nearly 3,000 vehicles and volunteers over the course of 2 years. From this data, the researchers:

- Identified events that could contain "meaningful interactions" between an automated vehicle and one driven by a human and created a simulation that replaced all the uneventful miles with these meaningful interactions.
- Programmed their simulation to consider human drivers the major threat to automated vehicles and placed human drivers randomly throughout.
- Conducted mathematical tests to assess the risk and probability of certain outcomes, including crashes, injuries, and near misses.
- Interpreted the accelerated test results using a technique called "importance sampling" to learn how the automated vehicle would perform, statistically, in everyday driving situations.

The accelerated analysis research was conducted on the two most common situations resulting in serious crashes. The first was where the automated vehicle was following one driven by a human, where adjustments must be made constantly for movements of the lead vehicle, as well as speed, road, and weather conditions, and other rapidly changing factors. The second involved a human-driven car cutting in front of the automated car that was being followed, in turn, by another human-driven vehicle. Three metrics—crash, injury, and conflict rates—were calculated along with the likelihood that one or more passengers in the automated vehicle would suffer moderate to fatal injuries. The accuracy of the evaluation was determined by conducting and then comparing accelerated and real-world simulations.

12.10 Photo-Realistic Simulator

Researchers from the University of Maryland, Baidu Research, and the University of Hong Kong developed a photo-realistic simulation system for training and validating self-driving vehicles [14]. The new system is said to provide a richer, more authentic simulation than current systems that use game engines or high-fidelity computer graphics and mathematically rendered traffic patterns.

Their system, called augmented autonomous driving simulation (AADS), could make self-driving technology easier to evaluate in the lab while also ensuring more reliable safety before expensive road testing begins.

"This simulation allows testing the reliability and safety of automatic driving technology before it is deployed on real cars and tested it on the highways or city roads," said Dinesh Manocha, a professor of electrical and computer engineering at the University of Maryland Institute for Advanced Computer Studies.

To ensure safety, AVs must evaluate and respond to the driving environment without fail. Given the innumerable situations that a car might encounter on the road, an autonomous driving system requires hundreds of millions of miles worth of test drives under challenging conditions to demonstrate reliability.

Although test drives could take decades to accomplish on the road, preliminary evaluations could be conducted quickly, efficiently, and more safely by computer simulations that accurately represent the real world and model the behavior of surrounding objects. Current state-of-the-art simulation systems described in scientific literature fall short in portraying photo-realistic environments and presenting real-world traffic flow patterns or driver behaviors.

AADS is a data-driven system that more accurately represents the inputs a self-driving car would receive on the road. Self-driving cars rely on a perception module that receives and interprets information about the real world, and a navigation module that makes decisions, such as where to steer or whether to break or accelerate, based on the perception module.

In the real world, the perception module of a self-driving car typically receives input from cameras and lidar sensors, which use pulses of light to measure distances of surrounding. In current simulator technology, the perception module receives input from computer-generated imagery and mathematically modeled movement patterns for pedestrians, bicycles, and other cars. It is a relatively crude representation of the real world. It is also expensive and time-consuming to create, because computer-generated imagery models must be hand generated.

The AADS system combines photos, videos, and lidar point clouds—which are like 3D shape renderings—with real-world trajectory data for pedestrians, bicycles, and other cars. These trajectories can be used to predict the driving behavior and future positions of other vehicles or pedestrians on the road for safer navigation.

Manocha said they are rendering and simulating the real world visually using videos and photos. Additionally, they are capturing real behavior and patterns of movement. The way humans drive is not easy to capture by mathematical models and laws of physics. Therefore, Manocha's team extracted data about real trajectories from all the video that was available, and they modeled driving behaviors using social science methodologies. Manocha noted that "this data-driven approach has given us a much more realistic and beneficial traffic simulator."

The scientists had a long-standing challenge to overcome in using real video imagery and lidar data for their simulation: Every scene must respond to a self-driving car's movements, even though those movements may not have been captured by the original camera or lidar sensor. Whatever angle or viewpoint is not captured by a photo or video must be rendered or simulated using prediction methods. This is why simulation technology has always relied so heavily on computer-generated graphics and physics-based prediction techniques.

To overcome this challenge, the researchers developed technology that isolates the various components of a real-world street scene and renders them as individual elements that can be resynthesized to create a multitude of photo-realistic driving scenarios.

With AADS, vehicles and pedestrians can be lifted from one environment and placed into another with the proper lighting and movement patterns. Roads can be recreated with different levels of traffic. Multiple viewing angles of every scene provide more realistic perspectives during lane changes and turns. Additionally, advanced image processing technology enables smooth transitions and reduces distortion compared to other video simulation techniques. The image processing techniques are also used to extract trajectories and, thereby, model driver behaviors.

References

1. Department of Transportation, "Federal Automated Vehicles Policy in Roadway Safety: Accelerating the Next Revolution," Department of Transportation, September 2016.
2. Anderson, J.M.,Kaltra, N.,Stanley, K.D.,Sorenson, P.et al., Autonomous Vehicle Technology A Guide for Policymakers (Santa Monica, CA: Rand Corp., 2016), 1-214.
3. U.S. Energy Information Admin, "Autonomous Vehicles: Uncertainties and Energy Implications," U.S. Energy Information Admin, May 2018.
4. Kelechava, B., "SAE Levels of Driving Automation," ANSI, November 16, 2018.
5. Datum, A. and Lotze, K., "The Role of Battery Management Systems in Autonomous Drive," *My Business Future*, 2017.
6. Ghasemi, M. and Song, X., "Powertrain Energy Management for Autonomous Hybrid Electric Vehicles with Flexible Driveline Power Demand," IEEE Transactions on Control Systems Technology 27, no. 5 (2019): 2229-2236.
7. Gupta, A., "Five Challenges in Designing a Fully Autonomous System for Driverless Cars," *IIOT World*, 2017.
8. Stewart, J., "Self-Driving Cars Use Crazy Amounts of Power, and It's Becoming a Problem," Wired, February 6, 2018.
9. Franklin, D., "NVIDIA Jetson AGX Xavier Delivers 32 TeraOps for New Era of AI in Robotics," Nvidia, December 12, 2018.

10. Abuelsamid, S., "Intel Unveils Its Automated Driving Compute Challenger to Nvidia," Forbes, January 8, 2018.

11. Gordon, J., "Autonomous Vehicles Are Going to Be Electric," Electric Cars, September 10, 2018.

12. Monsor, M., Giagkiozis, I., Mills, S.R., Purshouse, R.C. et al., "Real-Time Improved Power Management for Autonomous Systems," IFAC Proceedings 47, no. 3, (2014): 2634-2639.

13. Peng, H. and McCarthy, R.L., "A Concept to Assess the Safety Performance of Highly Automated Vehicles," University of Michigan, January 2019.

14. Li, W., Pan, C.W., Zhang, R., Ren, J.P.et al., "AADS: Augmented Autonomous Driving Simulation Using Data-Driven Algorithms," Science Robotics 4, no. 28 (2019): eaaw0863.

CHAPTER 13

CAN Bus

The Control Area Network (CAN) Bus enables ECUs to communicate with each other without a host computer, which avoids extensive load on the main controller. It is a high performance, high reliability, advanced serial communication protocol that effectively supports distributed real-time control. CAN is well-suited for the automotive environment because it produces a large number of short messages with high reliability in a rugged environment. CAN is a serial communications bus developed by BOSCH for the automotive industry that replaces complex wiring harnesses with a two-wire bus (Figure 13.1) [1–3].

13.1 CAN Basics

CAN is based on the open systems interconnection (OSI) reference model that transfers data among nodes connected in a network [5, 6]. The OSI reference model defines a set of seven layers through which the data passes during communication between devices connected in a network. Each layer has its specific function that supports the layer above and below. CAN uses only two of the layers:

1. **Data link layer** that packages raw data into frames transferred from physical layer. This layer is responsible for transferring frames from one device to another without errors. After sending the frame it waits for the acknowledgment from the receiving device. The data link layer has two sublayers:
   - MAC (medium access control) layer that performs frame coding, error detection, signaling, serialization, and de-serialization.
   - LLC (logical link control) layer that provides multiplexing mechanisms that make it possible for several network protocols to coexist within a multipoint network and be transported over the same network medium.

FIGURE 13.1 Automotive cabling with and without CAN [4].

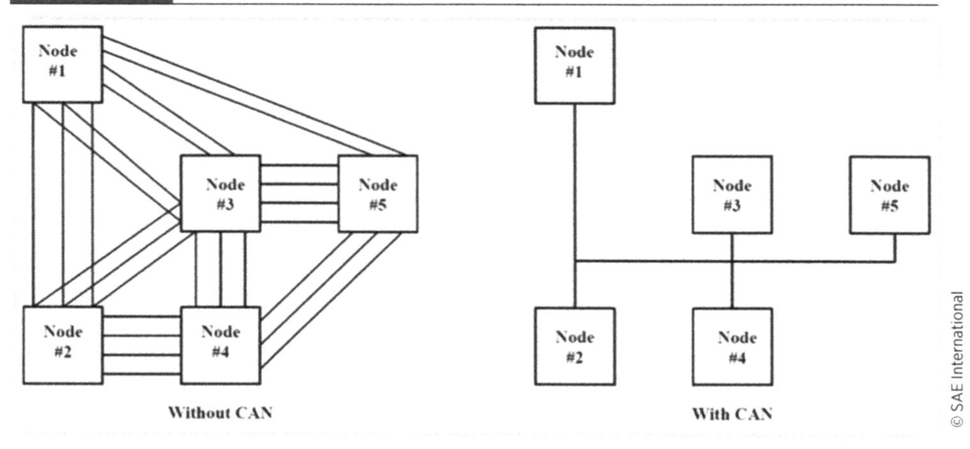

It performs the function of multiplexing protocols transmitted by MAC layer while transmitting and decoding when receiving and providing node-to-node flow and error control.

2. **Physical layer** transmits a bit from one device to another and regulates the transmission of bit streams. It defines the specific voltage and the type of cable to be used for transmission protocols. It provides the hardware means of sending and receiving data on carrier-defining cables, cards, and physical aspects.

The remaining five layers are communication layers that are left out by BOSCH's CAN specification for system designers to optimize and adapt according to their needs.

Characteristics of CAN Bus communication:

- Needs only two wires named CAN_H and CAN_L.
- Operates at data rates of up to 1 Megabit per second.
- Supports a maximum of 8 bytes per message frame.
- Does not support node IDs, only message IDs. One application can support multiple message IDs.
- Supports two message ID lengths, 11-bit (standard) and 29-bit (extended).
- Does not experience message collisions, which can occur under other serial technologies.
- Is not demanding in terms of cable requirements. Twisted-pair wiring is sufficient.

The CAN communications protocol, ISO-11898: 2003, describes communication on a network and conforms to the OSI model that is defined in terms of layers [4]. Actual communication between devices connected by the physical medium is defined by the physical layer of the model.

The CAN communication protocol is a carrier-sense, multiple-access protocol with collision detection and arbitration on message priority (CSMA/CD+AMP). CSMA means that each node on a bus must wait for a prescribed period of inactivity before attempting to send a message. CD+AMP means that collisions are resolved through a bit-wise arbitration based on a preprogrammed priority of each message in the identifier field of a message. The higher priority identifier always wins bus access. That is, the last logic-high in the identifier keeps on transmitting because it is the highest priority. Since every node on a bus takes part in writing every bit "as it is being written," an arbitrating node knows if it placed the logic-high bit on the bus [6].

The ISO-11898:2003 Standard, with the standard 11-bit identifier, provides signaling rates from 125 kbps to 1 Mbps (Table 13.1). The standard was later amended with the "extended" 29-bit identifier (Table 13.2).

The identifier allocates the priority CAN messages, which makes it attractive for a real-time control environment. The lower the binary message identifier number, the higher its priority. An identifier consisting entirely of zeros is the highest priority message on a network because it holds the bus dominant the longest. Therefore, if two nodes begin to transmit simultaneously, the node that sends a last identifier bit as a zero (dominant) while the other nodes send a one (recessive) retains control of the CAN bus and goes on to complete its message. A dominant bit always overwrites a recessive bit on a CAN bus.

A transmitting node constantly monitors each bit of its own transmission. The propagation delay of a signal in the internal loop from the driver input to the receiver output is typically used as a qualitative measure of a CAN transceiver. This propagation delay is referred to as the loop time (t_{LOOP}).

A CAN controller handles the arbitration process automatically. Because each node continuously monitors its own transmissions, as node B's recessive bit is overwritten by node C's higher priority dominant bit, B detects that the bus state does not match the bit that it transmitted. Therefore, node B halts transmission while node C continues on with its message. Another attempt to transmit the message is made by node B once the bus is released by node C. This functionality is part of the ISO 11898 physical signaling layer, which means that it is contained entirely within the CAN controller and is completely transparent to a CAN user [2].

The allocation of message priority is up to a system designer, but industry groups mutually agree on the significance of certain messages. For example, a manufacturer of motor drives may specify that message 0010 is a winding current feedback signal from a motor on a CAN network and that 0011 is the tachometer speed. Because 0010 has the lowest binary identifier, messages relating to current values always have a higher priority on the bus than those concerned with tachometer readings.

TABLE 13.1 Standard CAN with 11-bit identifier.

Bits	Name	Description
1	SOF	Start of frame
11	Identifier	Message priority
1	RTR	Remote transmission request
1	IDE	Identifier extension
1	r0	Reserved bit
4	DLC	Data length code
64	Data	Application data
16	CRC	Cyclic redundancy check
2	ACK	Acknowledge
7	EOF	End of frame
7	IFS	Interframe space

TABLE 13.2 Extended CAN with 29-bit identifier.

Bits	Name	Description
1	SOF	Start of frame
11	Identifier	Message priority
1	SRR	Substitute remote request
1	IDE	Identifier extension
18	Identifier	Message priority
1	RTR	Remote transmission request
1	r1	Reserve bit
1	r0	Reserve bit
4	DLC	Data length code
64	Data	Application data
16	CRC	Cyclic redundancy check
2	ACK	Acknowledge
7	EOF	End of frame
7	IFS	Interframe space

13.2 CAN Message Types

The four different message types, or frames, that a CAN bus can transmit are:

1. **The Data Frame** is the most common message type and comprises the arbitration field, the data field, the CRC field, and the acknowledgment field. The Arbitration Field contains an 11-bit identifier in the RTR bit which is dominant for data frames.
2. **Remote Frame's** intended purpose is to solicit the transmission of data from another node. The remote frame is similar to the data frame with two important

differences. First, this type of message is explicitly marked as a remote frame by a recessive RTR bit in the arbitration field. Second, there is no data.

3. **Error Frame** is a special message that violates the formatting rules of a CAN message. It is transmitted when a node detects an error in a message and causes all other nodes in the network to send an error frame as well. The original transmitter then automatically retransmits the message. An elaborate system of error counters in the CAN controller ensures that a node cannot tie up a bus by repeatedly transmitting error frames.
4. **Overload Frame** is mentioned for completeness. It is similar to the error frame with regard to the format, and it is transmitted by a node that becomes too busy. It is primarily used to provide an extra delay between messages.

A valid frame is a message considered to be error free when the last bit of the ending EOF field of a message is received in the error-free recessive state. A dominant bit in the EOF field causes the transmitter to repeat a transmission.

CAN signaling is differential, which provides robust noise immunity and fault tolerance. Balanced differential signaling reduces noise coupling and allows high signaling rates over twisted-pair cable. Balanced means that the current flowing in each signal line is equal but opposite in direction, resulting in a field-canceling effect that is a key to low noise emissions. The use of balanced differential receivers and twisted-pair cabling enhance the common mode rejection and high noise immunity of a CAN bus.

In a typical automotive system several ECUs are connected to the CAN_H and CAN_L lines, two are shown on Figure 13.2. Each ECU has a CAN controller and a CAN transceiver; and the CAN Bus lines are terminated in 120 Ω resistors. An integrated microcomputer accepts inputs from its dedicated CAN node and transfers data to the CAN controller. The CAN controller processes this data and relays in to the CAN transceiver. Also, the CAN controller receives data from the CAN transceiver, processes it and relays it to its integrated microcomputer.

The CAN transceiver is both a transmitter and a receiver. It accepts data from the CAN controller and converts it into a differential signal and sends it over the data lines. Additionally, the CAN transceiver receives data and converts it for the CAN controller [7].

In a typical data transfer CAN node 1 in ECU 1 sends data to its CAN controller. The CAN transceiver receives the data from the CAN controller, converts it into differential signals, and sends it to the network. All ECUs networked with the CAN data bus become receivers of the transmitted data. CAN node 2 in ECU2 checks to see whether

FIGURE 13.2 Typical CAN application [7].

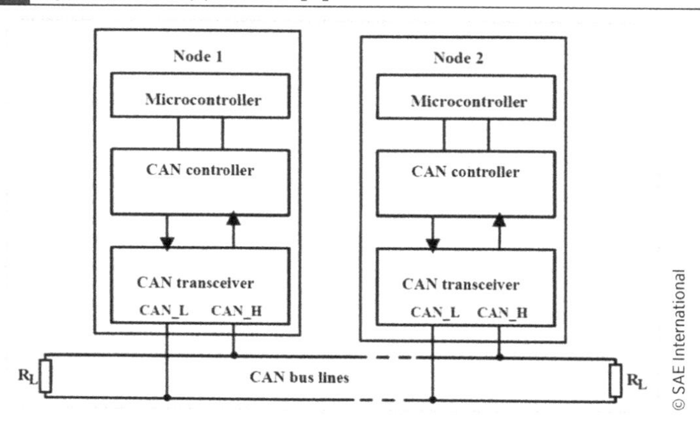

it requires the transmitted data for their functions. If the received data is important it is accepted and processed. If not, the received data is ignored.

13.3 FlexRay

FlexRay is an automotive network communications protocol developed by the FlexRay Consortium to govern onboard automotive computing [8]. It is designed to be faster and more reliable than CAN and TTP, but it is also more expensive. The FlexRay Consortium disbanded in 2009, but the FlexRay standard is now a set of ISO standards, ISO 17458-1 to 17458-5.

FlexRay supports high data rates, up to 10 Mbits/s, explicitly supports both star and "party line" bus topologies, and can have two independent data channels for fault-tolerance (communication can continue with reduced bandwidth if one channel is inoperative). The bus operates on a time cycle divided into two parts: the static segment and the dynamic segment. The static segment is pre-allocated into slices for individual communication types, providing stronger determinism than its predecessor CAN. The dynamic segment operates more like CAN, with nodes taking control of the bus as available, allowing event-triggered behavior.

13.4 Local Interconnect Network

LIN (Local Interconnect Network) is a serial network protocol used for communication between components in vehicles [9]. The need for a cheap serial network arose as the technologies and the facilities implemented in the car grew, while the CAN bus was too expensive to implement for every component in the car. European car manufacturers started using different serial communication technologies that led to compatibility problems.

In the late 1990s, the LIN Consortium was founded by five automakers (BMW, Volkswagen Group, Audi Group, Volvo Cars, Mercedes-Benz), with the technologies supplied (networking and hardware expertise) from Volcano Automotive Group and Motorola. The first fully implemented version of the new LIN specification (LIN version 1.3) was published in November 2002. In September 2003, version 2.0 was introduced to expand capabilities and make provisions for additional diagnostics features. LIN may be used also over the vehicle's battery power line with a special LIN over DC power line (DC-LIN) transceiver.

LIN over DC power line (DC-LIN) is being standardized as ISO/AWI 17987-8. CAN in automation has been appointed by the ISO Technical Management Board (TMB) as the Registration Authority for the LIN Supplier ID standardized in the ISO 17987 series.

All messages are initiated by the master with at most one slave replying to a given message identifier. The master node can also act as a slave by replying to its own messages. Because all communications are initiated by the master it is not necessary to implement a collision detection.

Current uses combine the low-cost efficiency of LIN and simple sensors to create small networks. These subsystems can be connected by backbone network (i.e., CAN in cars).

The LIN bus is an inexpensive serial communications protocol, which effectively supports remote application within a car's network. It is particularly intended for mechatronic nodes in distributed automotive applications but is equally suited to industrial applications. It is intended to complement the existing CAN network leading to hierarchical networks within cars.

References

1. Wikipedia, “CAN Bus.”
2. CSS Electronics, “A Simple Intro to CAN Bus,” CSS Electronics.
3. BOSCH, “CAN Specification Version 2.0,” BOSCH, July 1, 1996, 1-72.
4. Optimizel Softing, “CAN FD ISO 22898-1,” Optimizel Softing.
5. Texas Instruments, “Introduction to the Controller Area Network (CAN),” Texas Instruments, July 5, 2016, 1-17.
6. Hunting, B., “CAN Bus: Understanding the Basics,” NAPA Know How Blog, February 16, 2017.
7. Texas Instruments, “2N65HVD23x 3.3V CAN Bus Transceiver,” Texas Instruments, April 2018, 1-43 pages.
8. Wikipedia, “Flexray.”
9. Wikipedia, “Local Interconnect Network (LIN).”

CHAPTER 14

AEC Standards

The Automotive Electronics Council (AEC) is a US organization that sets qualification standards for components used in the automotive electronics industry. AEC was formed in 1993 by Chrysler, Ford Motor, and Delco Electronics to standardize the way components are qualified for automotive applications. The resulting AEC standards set requirements covering qualification definitions [1, 2].

14.1 AEC Standards Basics

AEC standards are based mainly on typical mission profiles [3, 4]. It is less effective with the untypical and stringent conditions increasingly common in vehicular applications. It is also a "one-time" qualification that typically takes place at the end of the development cycle. Thus, it takes no account of long-term production process stability [5].

AEC developed several specifications for automotive electronics:

- AEC-Q100: "Stress Test Qualification for Integrated Circuits" (1994)
- AEC-Q101: Discrete semiconductors
- AEC-Q102: Optoelectronic components
- AEC-Q104: Multichip modules
- AEC-Q200: Passive components

The parts are assigned Grades 0–3 based on their operating temperature (Table 14.1).

TABLE 14.1 Operating temperature categories.

Grade	Temperature range
0	-50°C to +150°C
1	-40°C to +125°C
2	-40°C to +105°C
3	-40°C to +85°C

14.2 AEC-Q100

AEC-Q100 is a set of integrated circuit qualification test sequences that establish common part qualification and quality system standards for ICs used in automotive electronics. It is the industry standard specification that outlines the recommended qualification requirements and procedures for packaged ICs embedded into automotive applications (Table 14.2).

14.3 AEC-Q101

AEC-Q101 defines minimum stress test-driven qualification requirements and reference test conditions for qualification of discrete semiconductors like transistors, diodes, and so on (Table 14.3). This document does not relieve the supplier of their responsibility to meet their own company's internal qualification program. Additionally, this document does not relieve the supplier from meeting any user requirements outside the scope of

TABLE 14.2 AEC-Q100 device tests.

AEC-Q100-001	Wire Bond Shear
AEC-Q100-002	Human Body Model (HBM) Electrostatic Discharge (ESD)
AEC-Q100-003	Decommissioned
AEC-Q100-004	IC Latch-Up
AEC-Q100-005	Nonvolatile Memory Write/Erase Endurance, Data Retention, and Operational Life
AEC-Q100-006	Decommissioned
AEC-Q100-007	Fault Simulation and Test Grading
AEC-Q100-008	Early Life Failure Rate (ELFR)
AEC-Q100-009	Electrical Distribution Assessment
AEC-Q100-010	Solder Ball Shear
AEC-Q100-011	Charged Device Model (CDM) Electrostatic Discharge (ESD)
AEC-Q100-012	Short-Circuit Reliability Characterization of Smart Power Devices For 12V Systems

TABLE 14.3 Discrete semiconductor tests.

AEC-Q101-001	Electrostatic Discharge Test—Human Body Model
AEC-Q101-002	Electrostatic Discharge Test—Machine Model
AEC-Q101-003	Wire Bond Shear
AEC-Q101-004	Miscellaneous Test Methods
AEC-Q101-005	Electrostatic Discharge Test—Capacitive Discharge Model

this document. In this document, "user" is defined as any company developing or using a discrete semiconductor part in production. The user is responsible to confirm and validate all qualification and assessment data that substantiates conformance to this document.

14.4 AEC-Q200

The AEC-Q200 qualification is the global standard for stress resistance that all passive electronic components must meet if they are intended for use within the automotive industry. Parts are deemed to be "AEC-Q200 qualified" if they have passed the stringent suite of stress tests contained within the standard [6].

AEC-Q200 Rev. D splits the level of qualification required for different parts of the industry into five grades, numbered 0–4 (Table 14.4).

14.5 AEC-Q102

AEC-Q102, March 15, 2017, Initial Release

Failure Mechanism-Based Stress Test Qualification for Discrete Optoelectronic Semiconductors in Automotive Applications

AEC-Q102 defines the minimum stress test-driven qualification requirements and references test conditions for qualification of discrete optoelectronic semiconductors (e.g., light emitting diodes, photodiodes, and laser components in all exterior and interior automotive applications). It combines state-of-the-art qualification testing, documented in various norms (e.g., JEDEC, IEC, MILSTD) and manufacturer qualification standards.

The following documents from AEC-Q101 are respectively valid also for qualification of discrete optoelectronic semiconductors according to AEC-Q102:

- AEC-Q101-001: Electrostatic Discharge Test—Human Body Model
- AEC-Q101-003: Wire Bond Shear Test
- AEC-Q101-005: Electrostatic Discharge Test—Charged Device Model

14.6 AEC-Q104

AEC-Q104, September 14, 2017

Failure Mechanism-Based Stress Test Qualification for Multichip Modules

TABLE 14.4 Test temperatures for passive components.

Grade	Temperature range	Application
0	−50°C to +150°C	Any application throughout the automotive industry, regardless of location within the vehicle.
1	−40°C to +125°C	Most underhood applications
2	−40°C to +105°C	Hot spots within the passenger compartments
3	−40°C to +85°C	Within most of the passenger compartment
4	0°C to +70°C	Nonautomotive use

AEC-Q104 contains a set of failure mechanism-based stress tests and defines the minimum stress test-driven qualification requirements and references test conditions for qualification of multichip modules (MCM). A single MCM consists of multiple electronic components enclosed in a single package that perform an electronic function. This document applies only to MCMs designed to be soldered directly to a printed circuit board assembly. MCM types not included in the scope of this document include:

Two assembly components or MCMs that a Tier 1/original equipment manufacturers (OEM) assembles onto a system.

- Light emitting diodes (LEDs), which are covered by AEC-Q102.
- Power MCMs may require specific considerations and qualification test procedures that are outside the scope of this document. A power MCM consists of multiple active power devices (i.e., IGBTs, power MOSFETs, diodes) and, if necessary, additional passive devices (e.g., temperature sensors, capacitors), which are integrated on a substrate.
- Solid-state drives (SSD).
- MCMs with exterior connectors that are not soldered to a board or other assembly.

For an MCM with embedded firmware, the firmware is considered an integral part of the MCM. As such, it is qualified as part of the overall system methodology, which is dependent on the type of MCM.

14.7 AEC Qualification

Automotive qualified parts are qualified by subjecting an appropriate sample of the parts to rigorous rounds of testing, for example:

- Stringent electrical testing, followed by stress, then a further round of testing to ensure the component's electrical integrity.
- The part's temperature sensitivity is determined by first exposing several samples for a prolonged period to the maximum temperature within their required testing range. Then, the samples are subjected to temperature cycling throughout their entire temperature range. Next, they are subjected to a further round of measurements to determine their temperature sensitivity.
- Exposing the part to a high degree of humidity for a prolonged period tests moisture resistance. The component's operational life is also checked to ensure it passes the required benchmark.
- Also tested is the component's resistance to solvents.
- Exposing components to high levels of g-force for prolonged periods of time tests mechanical shock and vibration resistance, and by cycling the parts through periods of vibration.
- Parts' solderability and resistance to soldering heat is checked to ensure they are fully operable, which involves exposing the components to high temperatures.
- The board flex and terminal strength of the components are also checked to ensure compliance with standard attachments.
- Finally, the parts undergo a strict visual inspection and a check to ensure their physical characteristics meet the required specifications.

14.8 Automotive Quality Standards

IATF 16949:2016 is the International Standard for Automotive Quality Management Systems. IATF 16949 was jointly developed by The International Automotive Task Force (IATF) members and submitted to the International Organization for Standardization (ISO) for approval and publication [7]. The standard is applicable to any organization that manufactures components, assemblies, and parts for supply to the automotive industry.

IATF 16949 emphasizes the development of a process-oriented, quality management system that provides for continual improvement, defect prevention, and reduction of variation and waste in the supply chain. The goal is to meet customer requirements efficiently and effectively.

14.9 Chinese Automotive Standards

China Automotive Technology and Research Center Co., Ltd. (CATARC), established in 1985, is a central government-level enterprise belonging to the State-owned Assets Supervision and Administration Commission of the State Council and a comprehensive science and technology corporate group with extensive influence in the automotive industry home and abroad [8]. Its mission is to promote the healthy and sustainable development of the automotive industry in China and strive to build CATARC into a more innovative first-class science and technology corporate group with comprehensive service capabilities and strong influence in the world.

CATARC businesses cover industrial service, standards, policy research, testing, engineering technology, R&D, certification, big data, engineering design and general contracting, consultation, NEV, commercialized and strategic rising businesses, and so on. CATARC is also active in automotive engineering research, auto safety technology research, auto energy saving technology research, and auto environmental protection technology research.

The organization developed the following standards for battery electric vehicles (BEVs) (Table 14.5) and fuel cell electric vehicles (FCEVs) (Table 14.6).

TABLE 14.5 China standard tests for battery electric vehicles.

BEV and key parts	
Serial number	**Description**
GB/T18384.1-2001	EV Safety Requirement Part 1: Energy Storage
GB/T18384.2-2001	EV Safety Requirement Part 2: Function and Protection
GB/T18384.3-2001	EV Safety Requirement Part 3: Electric Shock Protection
GB/T 4094.2-2005	Symbol of Operator, Indicator, and Signal of EV
GB/T 19596-2004	Electric Vehicle Terminology
GB/T 18385-2005	Electric Vehicle Power Performance Test Procedure
GB/T 18386-2005	Electric Vehicle Energy consumption and Range Test Procedure
GB/T 18387-2008	30 MHz EV EMC Limit and Test Procedure Broadband 9khz-30MHz
GB/T 18388-2005	Electric vehicle Type Approval Test Procedure
GB/T 24552-2009	EV Windshield Defrost and Defog Requirement and Test Procedure
GB/T 19836-2005	Electric Vehicle Instrument Panel
QC/T 838-2010	Electric Bus with Ultra-Capacitor

TABLE 14.6 China standard tests for fuel cell vehicles.

FCEV	
Serial number	**Description**
GB/T 24554-2009	Fuel Cell Engine Performance Test Procedure
GB/T 24549-2009	Fuel Cell Electric Vehicle Safety Requirement
GB/T 24548-2009	Fuel Cell Electric Vehicle Terminology
QC/T 816-2009	Specification of Hydrogen Refueling Vehicle
GB/T 26990-2011	Fuel Cell Electric Vehicles—Onboard Hydrogen System Specifications
GB/T 26991-2011	Fuel Cell Electric Vehicles—Maximum Speed Test Method

References

1. Electrical Engineering, "Is it Mandatory to Use AEC Qualified Components in Automotive Applications?," Electrical Engineering, January 2017.
2. Teschler, L., "Tougher Stress Tests for Automotive MOSFETs," Power Electronic Tips, November 17, 2017.
3. SAE, "What Do SAE Standards Offer the Automotive Electric Industry," SAE, January 2009, 1-12.
4. Wikipedia, "Automotive Electronics Council."
5. IHS Markit, Automotive Electronics Council (AEC) Publications, IHS Markit Standards Store.
6. Lawrie, E., "The AEC-Q200 Standard: What It Really Means," Golledge Electronics Ltd., October 6, 2015.
7. 16949 Academy, "What Is IATF16949?," 16949 Academy 2016.
8. CATARC, "Status of EV Standards in China," CATARC, May 18, 2012, 1-11.

about the author

Sam Davis has had 18 years of experience in electronic engineering design and management, 6 years in public relations, and 30 years as a trade press editor. He holds a BSEE from Case Western Reserve University and completed graduate work at the same school and UCLA.

He has worked as the editor-in-chief for PCIM (Power Electronics Technology) from 1984 to 2004 and powerelectronics.com from 2008 to 2019. Prior to that, he served as a field editor for CMP Publications' *EE Times* and *Electronic Buyer's News* as well as Cahners Publishing's *EDN* and *Electronic Business*.

Mr. Davis' technical writing experience has encompassed analog and digital technology, ranging from computer design to power supply design. Articles he has written also include semiconductor subjects, such as power semiconductors, hybrid circuits, and integrated circuits. He has also written articles on electric and autonomous vehicles, photovoltaic systems, drones, robots, and software. His articles have appeared in trade publications, such as *PCIM*, *Power Electronics Technology*, powerelectronics.com, *Real Time Computing*, *COTS Journal*, *Computer Design*, *EDN*, *Electronic Business*, *EE Times*, *Electronic Buyers News*, *Electronic Design*, and *Microcomputing*.

His engineering experience includes circuit and system design for Litton Systems, Bunker-Ramo, Rocketdyne, and Clevite Corporation. Design tasks included analog circuits, display systems, power supplies, underwater ordnance systems, and test systems. He also served as a program manager for a Litton Systems' Navy program.

Sam Davis is the author of *Computer Data Displays*, a book published by Prentice-Hall in the United States and Japan in 1969. He is also a recipient of the Jesse Neal Award for trade press editorial excellence. He also authored the 25-chapter *Power Management* ebook that appeared in powerelectronics.com.

He has one patent for naval ship construction that simplifies electronic system integration.

Index

CPSIA information can be obtained
at www.ICGtesting.com
Printed in the USA
LVHW070615180123
737395LV00004B/18

9 781468 601442